METEOROLOGICAL MONOGRAPHS

VOLUME 26   AUGUST 1996   NUMBER 48

STOCHASTIC LAGRANGIAN MODELS OF TURBULENT DIFFUSION

Howard C. Rodean

Atmospheric Science Division
Lawrence Livermore National Laboratory
Livermore, California

American Meteorological Society
45 Beacon Street, Boston, Massachusetts 02108

ISBN 1-878220-23-3
ISSN 0065-9401

Published by the American Meteorological Society
45 Beacon Street, Boston, MA 02108

Printed in the United States of America
by Braun-Brumfield, Inc., A Sheridan Group Company

TABLE OF CONTENTS

Foreword v

Preface vii

Chapter 1. Introduction 1

Chapter 2. Historical Review 3
    2.1 Brownian motion 3
    2.2 Turbulent diffusion 4

Chapter 3. Applicable Physics of Turbulence 7
    3.1 Reinterpretation of the Langevin equation for turbulent diffusion 7
    3.2 Kolmogorov's similarity hypotheses 8
    3.3 Lagrangian structure function 8
    3.4 Lagrangian and Eulerian statistics 9
    3.5 A solution for stationary homogeneous turbulence 9
    3.6 An alternate to the integral Lagrangian timescale 10

Chapter 4. Definitions for Stochastic Diffusion 13
    4.1 Random, stochastic, and Markov processes 13
    4.2 White noise and the Wiener process 14
    4.3 Realizable probability distribution functions 18

Chapter 5. The Fokker–Planck Equation 19
    5.1 Mathematical properties 19
    5.2 Derivation 20
    5.3 The forward and backward equations 23
    5.4 The Kramers–Moyal expansion 23

Chapter 6. Stochastic Differential Equations for Turbulent Diffusion 25
    6.1 Langevin equation for stationary, inhomogeneous, Gaussian turbulence . . . 25
    6.2 The Ito and Stratonovich interpretations 25
    6.3 The diffusion limit: The random displacement equation 28
    6.4 Applicability of the Langevin and random displacement equations 30

Chapter 7. Criteria for Stochastic Models of Turbulent Diffusion 31
    7.1 Descriptions of criteria 31
    7.2 Applying the well-mixed criterion 31

Chapter 8. Turbulent Diffusion in Three Dimensions 35
    8.1 Multivariate Gaussian distribution 35
    8.2 Applicable Langevin and Fokker–Planck equations 36
    8.3 Solution for the stochastic term in the Langevin equation 37
    8.4 Solutions for the deterministic term in the Langevin equation 37
    8.5 The Borgas solution and Thomson's "simplest" solution 40
    8.6 Solutions for the random displacement and diffusion equations 41

Chapter 9. Applications of Thomson's "Simplest" Solution 43
    9.1 Simplification for three dimensions 43
    9.2 Simplification to two dimensions and one dimension 44

Chapter 10. Application to the Convective Boundary Layer 45
    10.1 Approximating non-Gaussian turbulence 45
    10.2 Pioneering applications 47
    10.3 Model of Luhar–Britter and Sawford–Weil 47
    10.4 Four-moment model of Du, Wilson, and Yee 49
    10.5 Closure of Luhar, Hibberd, and Hurley 49

Chapter 11. The Boundary Condition Problem 51
    11.1 Historical review 51
    11.2 Study by Wilson and Flesch 52
    11.3 The random displacement model 54
    11.4 Appendix: A reflection algorithm 54
Chapter 12. Parameterization of Turbulence Statistics for Model Inputs 59
    12.1 Basic scaling parameters 59
    12.2 Vertical velocity variance 61
    12.3 Dissipation of turbulence kinetic energy 62
    12.4 Universal constant for the Lagrangian structure function 63
    12.5 Skewness of the vertical velocity 63
    12.6 Horizontal velocity variances 63
    12.7 Shear stress 64
    12.8 Unfinished business 64
Chapter 13. Brief Review of Four Topics 65
    13.1 Two-particle dispersion 65
    13.2 Intermittency and turbulent diffusion 65
    13.3 Kolmogorov's refined hypotheses 66
    13.4 Fractional Brownian motion 67
Chapter 14. In Conclusion 69
Acknowledgments 69
References 71
Annotated Bibliography 77
Author Index 81
Subject Index 83

Foreword

In a sense this monograph represents a watershed in the modern stochastic theory of turbulent dispersion; it is a sign that the subject is now mature enough to be presented in book form. The theory has its classical origins through Taylor and more recently Obukhov. However, I think it is probably the ready accessibility of powerful computing resources that has seen the rapid recent development since about 1980. These computational advances have been accompanied by, and are based heavily upon, dramatic analytical and theoretical advances in understanding. Many of us working in the field have had to pick up the mathematical machinery of the theory of stochastic processes as we went along. Fortunately though, there have been some skilled mathematicians interested in the subject, particularly Paul Durbin and later David Thomson, who have been able to inject the necessary rigour and to guide the rest of us to the essentials necessary for a rigorous treatment. Nevertheless, relatively few researchers have managed to achieve a good understanding of the subtleties of the underlying mathematics and its connection to the physics of the problem. Now, with this monograph, Howard has provided a means, for those with the will, to follow the detail and the subtleties. Those who take up the challenge will certainly be rewarded.

Howard and I first started corresponding in June 1990. Not for the first time in his career, Howard had decided to come to grips with a new subject. He brought with him a fresh mind, inquisitive, enthusiastic and eager to understand. He raised many questions; some of fine detail, some very deep, but few trivial. For me he provided a stimulus and a challenge to explain many of the things I thought I knew and understood very well, only to find under Howard's questioning that my understanding was shallow and that I needed to work hard to provide the answers.

Now I have before me the product of Howard's endeavours. It carries with it a beautiful sense of the historical development of the subject, and draws together a very diverse range of material in a coherent way. It makes no claim to being encyclopedic; on the contrary Howard has been very selective, but in the material he has chosen has provided much more than the usual explanation and mathematical detail. Finally, I should note that each time I have seen a new version of the manuscript it has included some new material, sometimes before it has appeared in the literature. The subject may be mature, but it is certainly not dead, and this book will undoubtedly contribute to its continued growth. Congratulations Howard, it has been a pleasure to be involved in a small way with your project.

Brian Sawford

Preface

My purpose in writing this monograph is to give atmospheric scientists a basic understanding of the physical and mathematical foundations of stochastic Lagrangian models of turbulent diffusion. I assume that the reader is generally familiar with the statistics of turbulence throughout the atmospheric boundary layer over the full range of stability conditions. Therefore the introductory emphasis is on 1) the physics of turbulence that is approximated by the assumed mathematical processes and 2) the peculiar properties of stochastic differential equations. I use the adjective "peculiar" because stochastic differential equations are very different from classical differential equations. I try to be clear and explicit in my descriptions of both physics and mathematics. In the interest of clarity in the mathematical derivations, I often give more details than is possible in a journal article. If I don't take the reader step-by-step, I give general directions for the path that we follow. I include "boxes" with definitions of some terms. My emphasis is on the mathematical and physical fundamentals, not on implementation. However, I include a chapter on parameterizations of turbulence statistics that *could be used* as model inputs. The list of references is comprehensive but not encyclopedic. I cite only those publications that I found to be useful.

I began my study of the Langevin equation and its application to turbulent diffusion in early 1989. I soon realized that this equation has properties that were unfamiliar to me at that time. I correctly suspected that it is very easy to misuse this equation to model turbulent diffusion if one is ignorant of some mathematical subtleties. In the fall of 1989 I began to write a scientific diary as I learned about the Langevin and Fokker–Planck equations, white noise and the related Wiener process, model design criteria, boundary conditions, parameterizations of turbulence statistics for use as model inputs, and failures and successes in applications. I continued writing through the summer of 1993. The result was published as an informal document (Notes on the Langevin model for turbulent diffusion of "marked" particles, *Lawrence Livermore National Laboratory Report UCRL-ID-115869*, 1994).

This monograph is a complete rewriting and an expansion of my *Notes*. I thank my friend and colleague at the Laboratory, Paul Gudiksen, for suggesting several years ago that I write this book. I owe much to my other Laboratory colleagues for their continued encouragement, help, and support: Richard Cederwall, Marvin Dickerson, Donald Ermak, Rolf Lange, John Nasstrom, and Thomas Sullivan. I thank Erik Näslund, FOA, Umeå, Sweden; John Wilson, University of Alberta, Edmonton, Canada; John Wyngaard, Pennsylvania State University, University Park, Pennsylvania; Donald Lenschow and Jeffrey Weil, NCAR, Boulder, Colorado; and Ashok Luhar and Brian Sawford, CSIRO, Mordialloc, Australia for their critical and helpful comments. Brian has been both friend and mentor during these years.

Chapter 1

Introduction

Stochastic Lagrangian models are a versatile tool in modeling diffusion of passive scalars in the turbulent atmospheric boundary layer. Different names are given to these models; e.g., Langevin, Markov chain, Monte Carlo, random displacement, and random walk, with the last being the most commonly used. Markov chain, Monte Carlo, and random walk are generic names in that they can apply to both the Langevin and random displacement models. In the random displacement model, the particle trajectory is the vector sum of random incremental changes in displacement. In the Langevin model, stochastic incremental changes in Lagrangian velocity are integrated in time to define the particle trajectory in space. (We do not consider the random acceleration model in which stochastic incremental changes in acceleration are integrated to obtain the velocity and displacement.) The Langevin model is more versatile and generally applicable than the random displacement model.

In applications, the dispersion and diffusion of a passive scalar or tracer (e.g., a chemical species or radioactivity) is numerically simulated by calculating the Lagrangian trajectories of tens of thousands of "marked" particles. The species concentration at a specific location at a given time is linearly proportional to the number of particles per unit volume. These calculations require high-speed digital computers for real-time or near-real-time simulations. Therefore, the development and application of these models have paralleled the advances in computer technology and software.

We use the acronyms LEM and RDM; RDM for the random displacement model and LEM for the Langevin model because it is based on the Langevin equation. We define both models in terms of perturbations to particles in a mean flow in the atmosphere that is numerically simulated by other means. (The turbulent flow, as a whole, can be modeled stochastically, but we do not consider that case.) We assume constant-density (not necessarily incompressible) flow because it simplifies the mathematics. We define terms like Lagrangian, Markov, stochastic, etc., in the following chapters.

We begin with a brief history of the efforts to model two phenomena, Brownian motion and turbulent diffusion, by means of stochastic differential equations. We review the physical phenomena in turbulence [i.e., Kolmogorov's (1941) hypotheses] that are approximated by our models and the statistics of turbulence that we use as inputs. We define mathematical and statistical terms like Markov process, realizability, and probability distribution function. We study the LEM for one-dimensional diffusion in stationary, inhomogeneous turbulence and transform it into the RDM. We derive the Fokker–Planck equation and show how it is related to the Langevin equation. We review several design criteria for stochastic Lagrangian models of turbulent diffusion, with emphasis on the "well-mixed" criterion. We derive the LEM for nonstationary, inhomogeneous turbulence in three dimensions and explain why there is no unique solution in more than one dimension. We propose simplified forms of the LEM for use in two and three dimensions.

We derive the only LEM that meets the well-mixed criterion for one-dimensional diffusion in the non-Gaussian convective boundary layer. We discuss some problems encountered in extending this model to more than one dimension.

Our next subject is the proper formulation of boundary conditions for the LEM and RDM. We find that boundary conditions, until recently, have been mentioned only slightly, if at all, in most of the papers in the literature. We use a recent paper as the principal basis for our discussion.

Our last subject concerning the LEM and RDM is the parameterization of turbulence statistics for use as inputs. We review some datasets that are available, and we propose specific parameterizations for use on a trial basis. We conclude by discussing the serious lack of data for all regions in the dimensionless height versus atmospheric stability parameter space for the atmospheric boundary layer.

We close with a brief review of publications on two-particle Lagrangian dispersion (we considered only the one-particle case in the above), the effect of intermittency on turbulent dispersion, Kolmogorov's refined (1962) hypotheses, and fractional Brownian motion.

LEM: Langevin equation model for turbulent diffusion of a scalar.

RDM: Random displacement model for turbulent diffusion of a scalar.

Chapter 2

Historical Review

2.1 Brownian motion

We begin with the following summary based on Chandrasekhar (1943), Doob (1953), Gardiner (1983, 1990), and van Kampen (1981, 1992). Our story begins in 1826, 1827, or 1828 (references differ in the date) when Robert Brown studied what became known as Brownian motion. He first observed tiny pollen grains suspended in water that exhibited very agitated and irregular motion. He was a botanist so it was natural for him to wonder if this motion was characteristic of only organic life. He continued his work with suspensions of other kinds of fine particles, organic and inorganic, and observed the same phenomenon. He concluded that this motion is not organic in origin.

Einstein published the first satisfactory explanation of Brownian motion in 1905; this work is considered to mark the beginning of stochastic modeling of natural phenomena. He considered Brownian motion as a random-walk process, as did Smoluchowski in 1906 and later years. Markov published a definitive work on the random walk in 1912.

Langevin published an alternate solution for Brownian motion in 1908 in which he introduced the equation that is named after him. He wrote it in the form

$$du/dt = -a_1 u + b\xi(t), \qquad (2.1)$$

where u is particle velocity, t is time, and a_1 is a damping coefficient associated with viscous drag on the particle. The product of the coefficient b and the random function $\xi(t)$ is a rapidly fluctuating acceleration component from irregular and unsymmetrical molecular bombardment on the particle. In subsequent studies of Brownian motion in electrical and gravitational fields, an additional acceleration term a_0 was added to the basic Langevin equation (Uhlenbeck and Ornstein 1930; Wang and Uhlenbeck 1945; Chandrasekhar 1943):

$$du/dt = a_0 - a_1 u + b\xi(t). \qquad (2.2)$$

We emphasize that the form in which Eqs. (2.1) and (2.2) are written is *primitive* because they are written in terms of the random function $\xi(t)$. The Langevin equation of 1908 was the first example of a stochastic differential equation, but an adequate mathematical basis for it was not available until more than 40 years later when Ito formulated his concepts for stochastic differential equations. This is because $\xi(t)$ is a generalized function and has only statistically defined properties. We define $\xi(t)$ and describe its properties in chapter 4. We present an intermediate form of the Langevin equation in chapter 3 and the *modern* form in chapter 5.

The Langevin equation is a *Lagrangian* equation whereas the next equation we consider is *Eulerian*. We interrupt our narrative to present the following definitions:

Lagrangian: An adjective identifying a system of coordinates based on the position (x, y, z) of a particle at time t relative to its position (a, b, c) at a reference time t_0. Lagrangian coordinates are physically more natural for describing fluid motion, but are more awkward to use.

Eulerian: An adjective identifying a system of coordinates fixed in space. The properties of a fluid such as density, temperature, and velocity are specified at a specific point in space (x, y, z) for a given time t. Eulerian coordinates are easier to use than Lagrangian coordinates.

The *Eulerian* equivalent to the Langevin equation is a partial differential equation, the Fokker–Planck equation. It is named after Fokker who published a paper on Brownian motion in 1914 and Planck who published a more general discussion in 1917. It is also called the "Smoluchowski equation" because of work he published in 1915–1916, but another related equation is more commonly identified by his name. Kolmogorov (1931) established a rigorous basis for the Fokker–Planck equation, so his name is applied in some references to two variations, the "forward" and "backward" equations. We write our introductory version of the Fokker–Planck equation as

$$\frac{\partial P(x,u,t)}{\partial t} + u\frac{\partial P(x,u,t)}{\partial x} = \\ -\frac{\partial}{\partial u}[(a_0 - a_1 u)P(x,u,t)] \\ +\frac{1}{2}\frac{\partial^2}{\partial u^2}[b^2 P(x,u,t)].$$

(2.3)

Here a_0, a_1, and b have the same definitions as in Eq. (2.2), and $P(x,u,t) \equiv P(x,u,t|x_0,u_0,t_0)$ where $P(x,u,t|x_0,u_0,t_0)$ is the joint conditional or transition probability density of the location x and velocity u at time t, given that it was at x_0 and u_0 at t_0. We study the Fokker–Planck equation in detail in chapter 5.

The work on the Langevin and Fokker–Planck equations as models of Brownian motion continued, and more recent publications are worthy of study. Two of the best are Uhlenbeck and Ornstein (1930) and its sequel, Wang and Uhlenbeck (1945). The significance of the first is indicated by the term "Ornstein–Uhlenbeck process" (an identifier of the *stationary* Brownian motion process modeled by Eqs. (2.1)–(2.3) with a_0 and a_1 constant and $P(x,u,t)$ independent of time) that is used in the texts by Arnold (1974), Gardiner (1983, 1990), and van Kampen (1981, 1992). Finally, there is the detailed and comprehensive review by Chandrasekhar (1943).

2.2 Turbulent diffusion

Taylor (1921) was the first to apply the statistics of Lagrangian particle motion to the problem of turbulent diffusion. He did not make any explicit references to the earlier works of Einstein, Langevin, and others on Brownian motion, but he noted similarities between turbulent diffusion and "the drunkard's [random] walk." He introduced the idea of the velocity autocorrelation function and obtained two fundamental results: 1) the root-mean-square displacement in turbulent diffusion is initially proportional to time, and 2) at late times it is proportional to the square root of time.

Our specific interest in Langevin and Fokker–Planck equations has its origin in Obukhov's (1959) suggestion that the Fokker–Planck equation be used to model turbulent diffusion. His suggestion was met with skepticism as shown by the published discussion of his paper. However, this proposal led to a new line of development for turbulent diffusion models. Most workers did not use the Fokker–Planck equation as suggested by Obukhov but, implicitly if not explicitly, used the Langevin equation. The first theoretical papers on this subject were published in the 1960s, the first numerical simulations appeared in the 1970s, and there have been significant advances in both theory and applications since 1980.

The first to take up Obukhov's idea was Lin who published two papers in 1960 that were incorporated into chapter 7 of Lin and Reid (1963). Novikov (1963) followed, and Krasnoff and Peskin (1971) published what is largely a critique of Lin's model. These works concerned theory, not applications.

Hanna (1979) did not mention any of the above in showing that a finite-difference equivalent of the Langevin equation is approximately valid for Lagrangian and Eulerian wind speed observations in the atmospheric boundary layer. Gifford (1982) cited Hanna's results as well as the above authors in proposing that the Langevin equation be used to model horizontal diffusion in the atmosphere over diffusion times ranging from seconds to days. Gifford's proposal was criticized by Sawford (1984). He noted that the Langevin equation is applicable to three-dimensional turbulence whereas relative diffusion on a global scale is quasi-two-dimensional.

Early numerical simulations with the Langevin equation as in Eq. (2.1) did not give physically realistic results in nonuniform turbulence fields. Specifically, particle concentrations tended to increase in regions with low velocity variance in violation of the second law of thermodynamics. Wilson et al. (1981) were among the first to propose the addition of a "drift correction" term, analogous to a_0 in Eq. (2.2), as a corrective measure. Legg and Raupach (1982) followed with a different proposal for a_0. Each correction was an ad hoc addition, based on physical reasoning, to the Langevin equation. Despite the difference between these proposals, they gave similar results in systems with small gradients of velocity variance. Wilson from one team and Legg from the other joined Thomson (Wilson et al., 1983) to show why the two models could, under certain conditions, give similar solutions. They also demonstrated that a certain modification of the Wilson et al. (1981) model gave the best results of all in inhomogeneous turbulence. They remarked in closing, "Though we can offer no proof that [the third model] is correct, it seems worthy of investigation."

Thomson (1984) used moment-generating functions for the probability densities of the velocity fluctuations u and the random function $\xi(t)$ to derive the model of Wilson et al. (1983). Van Dop et al. (1985) transformed the Langevin equation into its Eulerian equivalent, the Fokker–Planck equation, and determined coefficients in the Langevin equation by relating the random walk model to the Eulerian equations for the conservation of mass and species. Sawford (1986) found that their results were equivalent to those of Thomson and demonstrated that "...only the Wilson et al. (1983) model can be realized to satisfy the second law of thermodynamics for Gaussian inhomogeneous turbulence."

Thomson (1987) is a classic landmark paper. He used more general forms of the Langevin and Fokker–Planck equations in evaluating several design criteria for stochastic Lagrangian models of turbulent diffusion. He proved the well-mixed criterion to be equivalent to or superior to each of the other criteria. He showed that the random forcing must be Gaussian if the stochastic velocity process is to be continuous. Finally, he stated (but did not demonstrate) that there is no unique solution for the drift correction in more than one dimension. Luhar and Britter (1989) and Sawford (Weil 1990) independently used the well-mixed criterion of Thomson to derive a very complex variant of Eq. (2.2) for vertical diffusion in the non-Gaussian turbulence of the convective boundary layer. There is little mention, if any, of boundary conditions in most papers, including those on applications. Wilson and Flesch (1993) addressed this problem in the context of Langevin equations that meet the well-mixed criterion. They also examined the subject of noninfinitesimal timesteps.

All the above references beginning with Lin and Reid (1963) concern the Langevin equation model (LEM) in one form or another. The random displacement model (RDM) is mentioned briefly in only a few; e.g., van Dop et al. (1985) and Thomson (1987) who discussed the stochastic differential equation for random displacements as the *diffusion equation limit* of the Langevin equation. This limit is illustrated by the transformation from the stochastic velocity process (the Langevin equation) to a stochastic position process (the random displacement equation) in Arnold (1974) section 8.3. The result of this transformation is an equation that is equivalent to the classical diffusion equation. The RDM was discussed in detail by Durbin (1983) and applied by Durbin (1980) and Durbin and Hunt (1980). The mathematical transformation from the LEM to the RDM was demonstrated by Durbin (1983, 1984), Boughton et al. (1987), and de Bass and Troen (1989). We derive the RDM from the LEM in chapter 6.

Well-mixed criterion: If a species of passive "marked particles" is initially mixed uniformly in position and velocity space in a turbulent flow (homogeneous or inhomogeneous), it will stay that way (Thomson 1987). In other words, if the concentration of a species is initially uniform in a flow, it will remain uniform if there are no sources or sinks for the species.

Chapter 3

Applicable Physics of Turbulence

3.1 Reinterpretation of the Langevin equation for turbulent diffusion

In chapter 2 we gave a physical interpretation of the terms in Eqs. (2.1) and (2.2) for Brownian motion. In this and following chapters, we use Eq. (2.2) as the basis of our study. We rewrite it in a form attributed to Wilson et al. (1981) by Sawford (1985) and Novikov (1986):

$$d\left(\frac{w}{\sigma_w}\right) = -\left(\frac{w}{\sigma_w}\right)\frac{dt}{\tau_L}$$
$$+ \left(\frac{\partial \sigma_w}{\partial z}\right)dt + \left(\frac{2}{\tau_L}\right)^{1/2}\xi(t)dt.$$

$$(3.1)$$

(Actually, Wilson et al. calculated trajectories in transformed coordinates normalized to length, time and velocity scales at a reference height.) Here w is the Lagrangian vertical velocity fluctuation, σ_w is the standard deviation of w, z is the vertical coordinate, and t is time. In addition, τ_L is the timescale for the Lagrangian velocity autocorrelation and $\xi(t)$ is a random function. We define $\xi(t)$ in section 3.5 and chapter 4, and τ_L in sections 3.4 and 3.6. The first two terms on the right in Eq. (3.1) are deterministic and the third is random. The first term is a measure of the "fading memory" of the velocity at an earlier time, a function of the Lagrangian timescale for velocity autocorrelation. The second is the "drift correction" to satisfy the well-mixed criterion and is a function of the standard deviation of the velocity w. The third term is the representation of acceleration increments, integrated over time dt, from random pressure fluctuations with short correlation times on the order of Kolmogorov's timescale (defined and discussed in the next section). This equation is written in terms of a dimensionless velocity (the velocity divided by its standard deviation). Novikov (1986) suggested that the use of nondimensional velocity as in Eq. (3.1) is advantageous in modeling diffusion in systems with boundaries.

We obtain an equivalent to Eq. (3.1) in terms of the velocity and its variance by means of simple algebra and differential calculus:

$$dw = -\frac{w}{\tau_L}dt$$
$$+ \frac{1}{2}\left(1 + \frac{w^2}{\sigma_w^2}\right)\frac{\partial \sigma_w^2}{\partial z}dt + \left(\frac{2\sigma_w^2}{\tau_L}\right)^{1/2}\xi(t)dt.$$

$$(3.2)$$

This form is commonly used in the literature. The simple changes in the first and third terms on the right result from multiplication by the standard deviation of the velocity. The changes to the second term, the drift correction, are more complex. The first component inside the parentheses is equal to the explicit drift correction from Eq. (3.1) multiplied by the standard deviation of the velocity. The second part is a result of differentiating the left side of Eq. (3.1), so it is a drift correction component that is implicit in Eq. (3.1).

We quote Chandrasekhar's (1943) comment on Eq. (2.1) as a model for Brownian motion that is equally applicable to Eqs. (3.1) and (3.2) for turbulent diffusion:

> "we should draw attention . . . to the very drastic nature of assumptions implicit in the very writing of an equation of the form of [Eqs. (3.1) and (3.2)]. For we have in reality supposed that we can divide the phenomenon into two parts, one [random] in which the discontinuity of the events taking place is essential while in the other [deterministic] it is trivial and can be ignored."

This is an excellent description of the nature of stochastic differential equations. Note that we have not defined the nature of the quantity $\xi(t)dt$ other than to state that it is random. There is another problem if the coefficient of this quantity, a function of the velocity variance and the Lagrangian velocity time scale, is not constant. Van Kampen (1981, p. 244; 1992, p. 230) stated that the fully nonlinear Langevin equation has no well-defined meaning without a rule for its integration. Equation (3.2) is nonlinear if $2\sigma_w^2/\tau_L$ is a function of z where $dz = wdt$. We discuss these and other problems in the next three chapters. Meanwhile, we continue our study of the physical basis for the Langevin model for turbulent diffusion.

We deliberately use "velocity" instead of "particle velocity" in the above discussions because we consider only *passive* scalars or tracers, so the "particle velocity" is everywhere equal to the "fluid velocity." We assume that the fluid "particles" are "marked" at the time t_s that they flow through the (Eulerian) source location x_s, y_s, z_s.

3.2 Kolmogorov's similarity hypotheses

An excellent summary of Kolmogorov's (1941) hypotheses for the local properties of turbulence at very high Reynolds numbers is given in Lin and Reid (1963) section 16. We paraphrase as follows:

1) Small-scale turbulence is locally isotropic whether the larger scales are isotropic or anisotropic.

2) Small-scale motions are, for the most part, controlled by the viscous forces and the rate of energy transfer from the larger eddies (the first hypothesis).

3) At the large-scale end of the small-scale eddy spectrum, turbulence statistics are independent of the viscosity and are a function of only the rate of turbulence kinetic energy dissipation by the viscous forces in the smaller eddies (the second hypothesis).

The "large-scale end" described in item 3 marks a transition from the energy-containing larger scales to the energy-dissipating smaller scales. These hypotheses are the subject of section 21.3 in Monin and Yaglom (1975). They stated that 1) the Kolmogorov hypotheses provide a satisfactory explanation of many properties of turbulent flows, 2) predictions based on them are in good agreement with experimental data, but 3) there is no rigorous proof for these hypotheses.

Kolmogorov's scales for length, velocity, and time (based on the first hypothesis) are given on p. 348 of Monin and Yaglom as

$$\eta_K = (\nu^3/\varepsilon)^{1/4},$$
$$u_K = (\nu\varepsilon)^{1/4}, \quad (3.3)$$
$$\tau_K = \eta_K/u_K = (\nu/\varepsilon)^{1/2},$$

for length, velocity, and time, respectively. Here ε is the mean rate of turbulence kinetic energy dissipation and ν is the kinematic viscosity. The Reynolds number based on the above length and viscosity scales is

$$\text{Re} = \eta_K u_K/\nu = 1. \quad (3.4)$$

The inertial forces are therefore equal to the viscous forces at this scale. Sawford (1984, 1985) and others stated that the Langevin model for turbulent diffusion is applicable to high Reynolds number, three-dimensional turbulence because the particle velocity is correlated but the acceleration is not for time steps in the inertial subrange $\tau_K \ll \Delta t \ll \tau_L$.

3.3 Lagrangian structure function

We next consider Kolmogorov's second hypothesis that the turbulent statistics in the inertial subrange $\tau_K \ll \Delta t \ll \tau_L$ are independent of the viscosity ν and are a function of only the mean rate of turbulence kinetic energy dissipation ε. The Lagrangian structure function is defined as the ensemble average of the square of the change in the Lagrangian velocity of a particle in the time interval Δt (Monin and Yaglom 1975, p. 83),

$$D(\Delta t) = \langle [w(t+\Delta t) - w(t)]^2 \rangle = \langle (\Delta w)^2 \rangle. \quad (3.5)$$

If Δt is within Kolmogorov's inertial subrange, that is, $\tau_K \ll \Delta t \ll \tau_L$, (see Monin and Yaglom 1975, section 21.4, especially p. 359),

$$D(\Delta t) = C_0 \varepsilon \Delta t, \quad (3.6)$$

where C_0 is a universal constant and ε, as it is in Eq. (3.3), is the mean rate of turbulence kinetic energy dissipation. Consistency of the Langevin equation with Kolmogorov's theory requires that

$$\langle (\Delta w)^2 \rangle = C_0 \varepsilon \Delta t. \quad (3.7)$$

There is uncertainty about the value of this universal constant. Thomson (1987) and Borgas and Sawford (1991, 1994a) cited a value of $C_0 = 4 \pm 2$, based on evaluation of experimental data by Hanna (1981). Sawford (1991) found that numerical simulations suggest a value of 7. Rodean (1991) surveyed the literature and found estimates that ranged from $C_0 = 1.6 \pm 0.6$ to 2.1 to 2–5 to 5–10. He derived a semianalytical value of $C_0 = 5.7$. Pope (1994) concluded that C_0 is in the range 5.0–6.5. Du et al. (1995) recommended a value of 3.0 ± 0.5. We introduce the quantity $(2\sigma_w^2/C_0\varepsilon)$ as an alternate to the Lagrangian velocity integral timescale τ_L in section 3.6—after we have specified some properties of the random function $\xi(t)dt$.

3.4 Lagrangian and Eulerian statistics

We now address a problem that comes from using a Lagrangian model for turbulent diffusion when almost all the available data on turbulence statistics are Eulerian. Monin and Yaglom (1971) and McComb (1990) reviewed the Lagrangian description of turbulence in connection with turbulent diffusion. The latter reference is easier to read, and his section 12.2, "The problem of expressing the Lagrangian analysis in Eulerian coordinates," explicitly gives us our answer, originally presented by Lumley (1962). This work by Lumley is also cited in section 9.5 of Monin and Yaglom. Lumley's results are summarized as follows:

1) Any statistical function in Lagrangian coordinates can be represented by a power series in time of Eulerian quantities.

2) Eulerian homogeneity implies Lagrangian homogeneity and Eulerian isotropy implies Lagrangian isotropy.

3) The distribution of Lagrangian velocity is identical to that of Eulerian velocity in homogeneous turbulence (we assume the velocities of the passive scalar or tracer and the fluid are identical).

Specifically, our problem with respect to Eqs. (3.1) and (3.2) is as follows: What are the appropriate definitions of the velocity variance σ_w^2 and the Lagrangian velocity timescale τ_L?

In the case of Eq. (3.2), σ_w^2 is defined as the Eulerian velocity variance σ_{wE}^2. This definition was explicitly used in Wilson et al. (1983), Thomson (1984), and Sawford (1985). However, Legg and Raupach (1982) called σ_w^2 the Lagrangian velocity variance on p. 6 and assumed it was equal to the Eulerian variance on p. 9. Sawford (1986) stated that van Dop et al. (1985) made the substitution $\langle w^n \rangle = \langle w_E^n \rangle$ in their work. In Thomson (1987), σ_w^2 is explicitly defined as the Eulerian velocity variance. We gain an understanding in chapters 6 and 8 why this definition is correct when we follow Thomson (1987) and use the Fokker–Planck equation to derive the drift correction.

Taylor (1921) introduced the Lagrangian velocity correlation coefficient for homogeneous turbulence,

$$R_L(t - t_0) = \langle w(t)w(t_0) \rangle / \langle w^2(t_0) \rangle, \quad (3.8)$$

where t_0 is a reference time, and the integral timescale of the Lagrangian velocity,

$$\tau_L = \int_{t_0}^{\infty} R_L(t - t_0)dt. \quad (3.9)$$

He noted that the limiting form of his series expansion for the correlation cofficient is an exponential function,

$$R_L(t) = e^{-(t-t_0)/\tau_L}. \quad (3.10)$$

Finally, he derived the following relation for the diffusion coefficient,

$$K = \langle w^2(t_0) \rangle \tau_L. \quad (3.11)$$

The preceding equation is equivalent to

$$K = \sigma_w^2 \tau_L, \quad (3.12)$$

which we use in the RDM. We note that the "correlation coefficient" and "limiting form" given by Eqs. (3.8) and (3.10), respectively, are identified explicitly in Taylor's classic paper. On the other hand, the "integral timescale" in Eq. (3.9) and the "diffusion coefficient" in Eq. (3.11) are implicit in his equations but are not mentioned in the text. Instead of the Langevin equation, Taylor used an equivalent autoregression process in which the velocities at times t and $t + \Delta t$ were slightly uncorrelated. More explicit presentations of the above results are in Corrsin (1962), Lin and Reid (1963) sections 41–42, and Monin and Yaglom (1971) section 9.3.

3.5 A solution for stationary homogeneous turbulence

In the case of stationary homogeneous turbulence, Eq. (3.2) simplifies to

$$dw = -(w/\tau_L)dt + (2\sigma_w^2/\tau_L)^{1/2}\xi(t)dt. \quad (3.13)$$

This is related to the earliest form of the Langevin equation, Eq. (2.1), which we rewrite in differential form,

$$dw = -a_1 w dt + b\xi(t)dt, \quad (3.14)$$

where a_1 and b are constant. We use the derivation of Uhlenbeck and Ornstein (1930) for Brownian motion, which is given with variations in Arnold (1974) pp. 134–136 and van Kampen (1981) pp. 237–239. It was applied to turbulent diffusion by Lin and Reid (1963) section 45 and Legg and Raupach (1982).

The classical Riemann–Stieltjes calculus gives the solution of Eq. (3.14) as

$$w(t) = w(0)e^{-a_1 t} + be^{-a_1 t}\int_0^t e^{a_1 s}\xi(s)ds. \quad (3.15)$$

Before we can evaluate the integral, we must define some statistical properties of the velocity fluctuation $w(t)$ and the random process $\xi(t)$. Their ensemble averages are

$$\langle w(t)\rangle = 0 \quad (3.16a)$$

and

$$\langle \xi(t)\rangle = 0. \quad (3.16b)$$

The fluctuation $w(t)$ and the process $\xi(t)$ are uncorrelated so

$$\langle w(t)\xi(t)\rangle = 0. \quad (3.16c)$$

The covariance of $\xi(t)$ for times t and s is

$$\langle \xi(s)\xi(t)\rangle = \delta(t - s) \quad (3.16d)$$

where δ is the Dirac delta function with dimension t^{-1}. Uhlenbeck and Ornstein were not as drastic as the condition given by Eq. (3.16d); they assumed the equivalent of $\xi(s)$ and $\xi(t)$ being correlated only if $|t - s| \approx \tau_K$, the Kolmogorov timescale defined in Eq. (3.3). We square Eq. (3.15), take the ensemble average of the result, apply Eqs. (3.16a–d), and evaluate the resulting integral to get

$$\langle w^2(t)\rangle = \langle w^2(0)\rangle e^{-2a_1 t} + \frac{b^2}{2a_1}\left[1 - e^{-2a_1 t}\right]. \quad (3.17)$$

If the turbulence is stationary and homogeneous, we get from Eq. (3.17) and the work of Lumley (1962),

$$\sigma_w^2 = \langle w^2(t)\rangle = \langle w^2(0)\rangle = b^2/2a_1 = \sigma_{wE}^2. \quad (3.18)$$

If we multiply Eq. (3.15) by $w(0)$, take the ensemble average of the result, and apply Eq. (3.16c), we get the result,

$$\langle w(t)w(0)\rangle = \langle w^2(0)\rangle e^{-a_1 t}. \quad (3.19)$$

From Eqs. (3.8)–(3.10), we get the solution for a_1:

$$a_1 = \tau_L^{-1}. \quad (3.20)$$

This definition of τ_L as an *integral* timescale is valid *only* for stationary homogeneous turbulence. Finally, from Eqs. (3.18) and (3.20),

$$b^2 = 2\sigma_w^2/\tau_L. \quad (3.21)$$

Then Eq. (3.14) becomes identical to Eq. (3.13):

$$dw = -(w/\tau_L)dt + (2\sigma_w^2/\tau_L)^{1/2}\xi(t)dt. \quad (3.22)$$

It follows from the dimensions of velocity, the Lagrangian velocity integral timescale, and the velocity variance in Eq. (3.22) that the dimension of the random process $\xi(t)$ is $t^{-1/2}$.

This is as far as we can go with the classical Riemann–Stieltjes calculus. It is clear that we must study further the properties of the quantity $\xi(t)dt$ and the calculus for stochastic differential equations in order to obtain rigorous solutions of Eqs. (3.1) or (3.2) for inhomogeneous turbulence.

3.6 An alternate to the Lagrangian velocity integral timescale

We can obtain more information from the preceding section. We use the square of Eq. (3.14), apply Eqs. (3.16a–d), and anticipate some results concerning $\xi(t)dt$ in chapter 4 to get

$$\langle (dw)^2\rangle = b^2 dt. \quad (3.23)$$

Then from Eqs. (3.7) and (3.23),

$$b = (C_0\varepsilon)^{1/2} \quad (3.24)$$

so we get an alternate definition for the Lagrangian velocity timescale,

$$\tau_L = 2\sigma_w^2/C_0\varepsilon. \quad (3.25)$$

We use this result to rewrite Eq. (3.22) as

$$dw = -(C_0\varepsilon/2\sigma_w^2)wdt + (C_0\varepsilon)^{1/2}\xi(t)dt. \quad (3.26)$$

It appears that Pope (1983) was the first to use $(C_0\varepsilon)^{1/2}$ as the coefficient of $\xi(t)dt$ in writing the Navier–Stokes equation for the incremental change of Lagrangian velocity as the sum of mean and fluctuating components. Van Dop et al. (1985) were the first to use the equivalent in the Langevin model for turbulent diffusion (their $2\pi B = C_0$). Thomson (1987) and others have followed this practice, so it has become standard.

There are major advantages in using Eq. (3.26) instead of Eq. (3.22). Let us consider the parameters in these equations: the variance of the velocity w, σ_w^2; the Lagrangian velocity time scale, τ_L; the mean rate of dissipation of turbulence kinetic energy, ε; and the universal constant for the Lagrangian structure function, C_0. Of these parameters, σ_w^2 is directly and readily measured, it is more difficult to determine ε from the turbulence kinetic energy budget equation, and it is extremely difficult to measure τ_L in Lagrangian coordinates. As a result, parameterizations for σ_w^2 and ε are available, but those for τ_L are not. As we have noted, τ_L has an *exact* definition as an integral timescale *only* in homogeneous turbulence. It has a different interpretation in inhomogeneous turbulence: It is a local velocity decorrelation timescale. On the other hand, as stated in section 3.3, there is uncertainty about the value of C_0. From an intuitive point of view, it may be easier to think in terms of Eq. (3.22), but it is more rational to use Eq. (3.26) in numerical simulations.

Chapter 4

Definitions for Stochastic Diffusion

Here and in chapters 5 and 6, we frequently refer to the textbooks on stochastic processes by van Kampen and Gardiner. For convenience, reference is generally made to the earlier editions, Gardiner (1983) and van Kampen (1981), because we used them in much of our earlier work (Rodean 1994). There are exceptions, for example, to van Kampen (1992) in cases involving the Langevin equation and to Gardiner (1990) for discussion of the procedures of Ito and Stratonovich for integrating stochastic differential equations.

4.1 Random, stochastic, and Markov processes

We have used "random" and "stochastic" without explicitly defining these words. The following definitions are from the *The New Shorter Oxford English Dictionary* (1993):

> *Random* (adjective): "Governed by or involving equal chances for each of the actual or hypothetical members of a population; produced or obtained by a process of this kind (and therefore completely unpredictable in detail)."

> *Stochastic* (adjective): "Randomly determined; that follows some random probability distribution or pattern, so that its behavior may be analyzed statistically but not predicted precisely."

These definitions suggest "random" and "stochastic" are synonymous, but there are implications that the former is more "physical" and the latter more "mathematical." This is confirmed by the opening sentence in the preface in Doob (1953): "A stochastic process is the mathematical abstraction of an empirical process whose development is governed by probabilistic laws."

Brownian motion was described by Wang and Uhlenbeck (1945) as a "Markov process." Turbulent diffusion was described as a Markov process by many workers beginning with Obukhov (1959) and Lin and Reid (1963) through Sawford (1985) and van Dop et al. (1985) to Thomson (1987) and beyond.

The index in Doob lists four stochastic processes that are of interest to us: Brownian motion, diffusion, Gaussian, and Markov. Brownian motion is a diffusion process that is both Gaussian and Markovian. Gaussian processes, as implied by the name, involve Gaussian random probability distributions. We study specific Gaussian processes in subsequent sections and chapters. We now define the Markov process.

In 1906 Markov formulated the process that was subsequently named after him. It can be defined in words as "the 'future' is independent of the 'past' when we know the 'present' " and "the past and future are statistically independent when the present is known" (Arnold 1974, pp. 27 and 29). Another definition was given by van Dop et al. (1985): "A stochastic process is Markovian when its future state depends only on its present state (and some transition rule)." We use mathematical equivalents of these words when we derive the Fokker–Planck equation in chapter 5. Doob (1953) discussed Markov processes in two chapters; chapter VI on *continuous* Markov processes is the one of interest to us. Other discussions of Markov processes are presented in Arnold (1974), the 1981 and 1992 editions of van Kampen, the 1983 and 1990 editions of Gardiner, and Øksendal (1992).

Markov process: A stochastic process that has zero memory of the past and a future that is a function of the present and some statistical rule for the transition.

Simulation of a Markov process: We illustrate the Markov process with the random walks presented in Fig. 4.1. We show four examples to emphasize the variability of such processes. Each step is one unit in length, and the total length of each walk is 256 steps. The directions of successive steps are independent. The direction of each step was selected from a population of pseudorandom numbers uniformly distributed over the range $0 < \theta < 2\pi$. Therefore, the "past" that led to a specific "present" could have been at a point anywhere on a circle centered at the "present" with a radius equal to the unit step. Similarly, the "future" could be anywhere on the same circle. [These figures were constructed with the "RandomWalk" package in *Mathematica* (Maeder 1991, p. 260).]

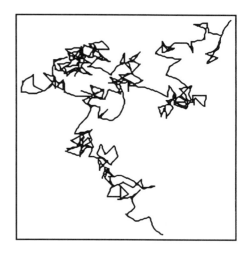

FIG. 4.1. Four random walks. In each walk, the direction of every one of the 256 equal-length steps is independent and in a uniformly random direction. Therefore, each walk is a Markov process.

4.2 White noise and the Wiener process

We stated in Eqs. (3.16a–d) that the ensemble average of the random function $\xi(t)$ is zero and its covariance for times s and t is the Dirac delta function.

These properties are noted in the Introduction of Arnold (1974) pp. xi–xii where $\xi(t)$ is also described as "white noise," a *hypothetical* stationary, Gaussian, stochastic process with constant spectral density on the real frequency axis. Arnold added that $\xi(t)$ is everywhere discontinuous and that its integral is a continuous, but not differentiable, pro-

cess. He made it clear that this process does not exist in the physical sense but is a very useful mathematical idealization.

The name white noise is given to $\xi(t)$ because it has a uniform spectral density, a characteristic equivalent to the Dirac delta correlation function. The Fourier transform of the covariance of $\xi(t)$, given by Eq. (3.16d), is

$$F(\omega) = \int_{-\infty}^{\infty} e^{i\omega t}\delta(t)dt = 1, \qquad (4.1)$$

where $i = \sqrt{-1}$ (Arnold 1974, pp. 50–51; Papoulis

1962, pp. 36–37). In a discussion of white noise, Gardiner (1983) pp. 18–20 showed that, the shorter the correlation time, the higher the frequency out to which the spectrum is flat. Van Kampen (1981) pp. 237–238 and 241-242 identified the above properties of white noise and stated, "Thus $[\xi(t)]$ may be visualized as a dense succession of small positive and negative pulses."

We introduce the "Wiener process," the continuous but not differentiable time integral of $\xi(t)$ as defined above. Doob (1942) resolved the problem of finding a proper stochastic interpretation of the Langevin equation by writing it in differential form in terms of the Wiener process, although he did not identify the latter as such [see his Eq. (3.3)]. Arnold (1974) and Gardiner (1983, 1990) emphasized the Wiener process because they considered it, as a continuous function, to be superior to discontinuous white noise for use in integrating stochastic differential equations. On the other hand, van Kampen (chapter IX, "The Langevin Approach," 1992) wrote every Langevin equation in terms of $d(\ )/dt$ and white noise as in our Eqs. (2.1) and (2.2), except at the bottom of p. 232 where he wrote a Langevin equation in terms of $d(\ )$, $(\ )dt$, and the Wiener process. Beginning with this chapter, we use the Wiener process instead of white noise. The following discussion, with some specific citations, is based partly on Arnold but mostly on Gardiner.

The Wiener process, $W(t)$, is related to white noise, $\xi(t)$, as follows:

$$W(t) = \int_0^t \xi(s)ds. \qquad (4.2)$$

Arnold (pp. xii and 53) indicated that the above can be written symbolically as

$$dW(t) = \xi(t)dt. \qquad (4.3a)$$

Gardiner (p. 82) presented the interpretation

$$dW(t) \equiv W(t+dt) - W(t) = \xi(t)dt, \qquad (4.3b)$$

where $dW(t)$ is an incremental Wiener process. The form $dW(t)/dt = \xi(t)$ should not be used because the process $\xi(t)$ is everywhere discontinuous; therefore, $dW(t)/dt$ is not defined. In other words, $W(t)$ is not differentiable.

The Wiener process is a continuous, Gaussian, Markov process that is a model of diffusion. Wiener (1923) studied it extensively, hence it is named after him. The following special form of the Fokker–Planck equation [Eq. (2.3) with the drift coefficient, a_0, and viscous damping, a_1, equal to zero; a diffusion coefficient, b^2, of unity; and only one variable,

$w(t)]$ is synonymous with (or defines) the Wiener process:

$$\frac{\partial}{\partial t}P(w,t|w_0,t_0) = \frac{1}{2}\frac{\partial^2}{\partial w^2}P(w,t|w_0,t_0), \qquad (4.4a)$$

where $P(w,t|w_0,t_0)$ is the conditional or transition probability density that a particle will be in state w at time t given that it was in state w_0 at time t_0. The initial condition for the conditional probability density is

$$P(w,t_0|w_0,t_0) = \delta(w-w_0), \qquad (4.4b)$$

where $\delta(w-w_0)$ is a Dirac delta function. The solution of Eqs. (4.4a,b) is the Gaussian distribution,

$$P(w,t|w_0,t_0) =$$
$$[2\pi(t-t_0)]^{-1/2}\exp[-(w-w_0)^2/2(t-t_0)].$$
$$(4.5)$$

The mean of the process $W(t)$ is

$$\langle W(t)\rangle = w_0 \qquad (4.6a)$$

and its variance is

$$\langle[W(t)-w_0]^2\rangle = t - t_0. \qquad (4.6b)$$

We use the analogs to Eqs. (4.6a,b) for the *incremental* Wiener process:

$$\langle dW(t)\rangle = 0, \qquad (4.7a)$$

and

$$\langle[dW(t)]^2\rangle = dt. \qquad (4.7b)$$

Equations (4.7a,b) are consistent with Eqs. (3.16a–d) and (3.23)–(3.26). Arnold (p. 46) stated and Gardiner (pp. 69–70) demonstrated that the increments $dW(t)$ are independent of each other and the initial value of $W(t)$, and that the distribution of $dW(t)$ is Gaussian.

There is a good discussion of the physical reality of *continuous Markov* processes on p. 45 of Gardiner. In summary, they do not exist in reality. Real processes have finite characteristic memory, as represented by the Lagrangian velocity timescale in the first term on the right in Eqs. (3.1) and (3.2). The random term (the last term) in these equations is a model of the forcing that has a characteristic memory with a duration on the order of the Kolmogorov timescale t_K defined by Eq. (3.3). However, Eqs. (3.1) and (3.2) may be regarded as Markovian for $\Delta t \gg t_K$. Continuous Markov processes do exist mathematically and are the subject of much literature [e.g., chapter VI in Doob (1953)].

Furthermore, Sawford and Borgas (1994) demonstrated that the property of continuity in a Markov model of turbulent diffusion is required for consistency with Kolmogorov's hypotheses discussed in section 3.2. We make use of the Gaussian property of both the input [random forcing associated with $dW(t)$] and the output (velocity fluctuations) in Langevin equations as models of the turbulent diffu-sion process. Van Kampen (1981, p. 88; 1992, p. 84) cited Doob (1942) in stating that the Ornstein–Uhlenbeck process [modeled by the Langevin equation in Uhlenbeck and Ornstein (1930)] is essentially the only process that is stationary, Gaussian, and Markovian. However, we must remember that Gaussian turbulence, for the most part, is only a useful fiction and approximation (e.g., Pope 1987).

White noise: A stationary, Gaussian, stochastic process $\xi(t)$ with a uniform spectral density that is everywhere discontinuous and whose time integral $W(t)$ is a continuous but not differentiable process.

Wiener process: A continuous, Gaussian, Markov process that is a model of diffusion. It is the continuous but not differentiable time integral of white noise. It may be called a continuous random (or drunkard's) walk in one dimension (Durbin 1983, p. 5). As a continuous function, it is superior to discontinuous white noise in integrating stochastic differential equations. The incremental Wiener process, $dW(t)$, is used in the integration process.

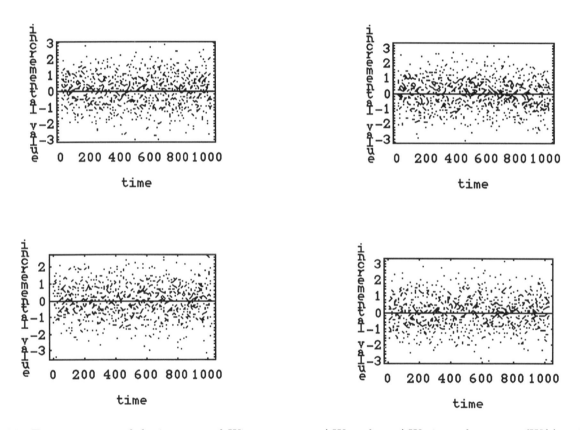

FIG. 4.2. Four sequences of the incremental Wiener process, ΔW_n, where ΔW_n is analogous to $dW(t)$ as in Eqs. (4.3a,b). The successive values of ΔW_n are independent, so they are plotted at distinct dots. Each set of 1024 dots is a different pseudorandom sequence of ΔW_n with a Gaussian distribution ($\langle \Delta W_n \rangle = 0, \langle \Delta W_n^2 \rangle = 1$) in analogy to Eqs. (4.7a,b) for $dW(t)$. The dots are spaced uniformly along the horizontal axis, with each space representing an increment of time, Δt. These sequences were summed to form the respective Wiener processes shown in Fig. 4.3.

Simulation of a Wiener process: *Mathematica* (Wolfram 1991, pp. 552 and 585) can be used to generate normal or Gaussian distributions. We learned in a series of numerical experiments that a sequence of at least 1000 pseudorandom numbers is required to obtain good approximations to the following theoretical values for Gaussian distributions: mean, 0; variance, 1; skewness, 0; and kurtosis, 3. Therefore, we used sets of 1024 pseudorandom numbers with Gaussian distributions in constructing the illustrations in Figs. 4.2 and 4.3. The successive values of Wiener process increments, ΔW_n, are independent, so we plotted the four sets of ΔW_n values in Fig. 4.2 as dots representing four sequences of random numbers. These number sequences were successively summed to obtain the corresponding Wiener processes in Fig. 4.3. This summation is analogous to the integration in Eq. (4.2); see the equation in the caption to Fig. 4.3. The results are jagged approximations to the continuous but not differentiable Wiener process. These Wiener processes are random walks in one dimension. Note that each process is significantly different from the others. Thousands of processes must therefore be calculated in order to obtain meaningful statistics; e.g., the mean concentration of "marked" particles diffusing in turbulent flow.

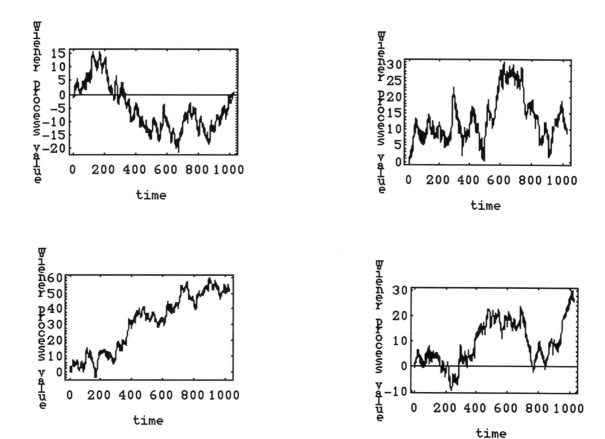

FIG. 4.3. Four Wiener processes, W_N, formed by summing the respective sequences of Wiener process increments, ΔW_n, shown in Fig. 4.2, where

$$W_N = \sum_{n=1}^{N} \Delta W_n. \qquad (4.8)$$

4.3 Realizable probability distribution functions

Let us begin this section with a definition from *The New Shorter Oxford English Dictionary* (1993):

> *Realization* (noun): [Statistics] "A particular series which might be generated by a specified random process." *Realizability* (noun) and *Realizable* (adjective) are briefly discussed under this entry.

With respect to "realizable" probability distribution functions, this definition means functions that have a real mathematical existence. The question of realizability arises in using non-Gaussian processes instead of the incremental Wiener process for the input (stochastic forcing) of Lagrangian models of diffusion. The following historical survey illustrates the necessity of proper stochastic model formulation in order to avoid nonrealizable probability distribution functions.

Thomson (1984) compared the performance of Langevin models with two drift corrections: one recommended by Wilson et al. (1983) and another proposed by Legg and Raupach (1982). He used moment-generating functions to solve for the moments of the stochastic forcing function (the input) required by each model, assuming a Gaussian distribution of the particle velocity fluctuations (the output). He determined that the model of Wilson et al. required a Gaussian input (the incremental Wiener process) but the Legg and Raupach model required a non-Gaussian input defined by a probability distribution function with nonzero values for the first three moments and zero values for all higher moments.

Thomson used three sets of numerical simulations to test 1) the Wilson et al. model with Gaussian forcing, 2) the Legg and Raupach model with Gaussian forcing [used in Legg and Raupach (1982)], and 3) the Legg and Raupach model with an *approximation* to the required non-Gaussian distribution. He found that, in moderately inhomogeneous turbulence, the Legg and Raupach model yielded better results with non-Gaussian forcing. In addition, the performance of the non-Gaussian Legg and Raupach model was comparable to that of the Gaussian Wilson et al. model in the moderate case, but the latter model was far superior in strongly inhomogeneous turbulence.

The reason for this difference between the performance of the (non-Gaussian) Legg and Raupach model and the (Gaussian) Wilson et al. model is that the *exact* non-Gaussian forcing function required by the Legg and Raupach model *is not realizable*. That is, it does not exist.

Van Dop *et al.* (1985) discussed Thomson's (1984) solution for a three-moment non-Gaussian forcing function in section 5 on skewed random acceleration. They noted that a number of difficulties are encountered in applying a skewed random forcing function, including the fact that differential equations of higher than second order (than that of the Fokker–Planck equation) cannot describe the evolution of a probability density function because the condition that the distribution $P(z, w, t)$ be positive everywhere is violated. They cited van Kampen (1981) who made similar comments and, in turn, cited Pawula (1967b). Pawula (1967a,b) studied differential equations for random processes of the second order (the Fokker–Planck equation) and higher (the Kramers–Moyal expansion that we discuss in section 5.4). He found that if such equations are to be of finite order, they must be of order two or less. Pawula and van Kampen explicitly agreed that, in practice, higher-than-second-order truncations are not necessarily useless, but should be used with caution.

Thomson recognized the problem of realizability in section 2.3 of his 1987 paper. He remarked that a Langevin model that requires non-Gaussian forcing must either be nonexistent or have discontinuous phase-space trajectories. He emphasized that the random forcing for a continuous process must be Gaussian. Doob (1942) stated in the text following his Eq. (3.14) that "... the only differential processes whose path functions $[dW(t)]$ do not have jumps have the property that $[dW(t)]$ is Gaussian. Then it is only in this case which will lead to the [Ornstein–Uhlenbeck] process, that $u(t)$ will not have jumps."

Sawford (1986) also discussed realizability. He stated that truncation of the Kramers–Moyal expansion at any higher order than two (the Fokker–Planck equation) results in nonphysical behavior such as negative P and cited Gardiner (1983) who, in turn, cited Marcienkiewicz (1939) who showed that either 1) all but the first two cumulants vanish or 2) there are an infinite number of nonvanishing cumulants. This means the Kramers–Moyal expansion must be truncated at $\partial^2 P/\partial w^2$ or not at all [see Eq. (4.4a)].

Lukacs expanded the work of Marcienkiewicz in a paper (1958) and a book (1960) section 7.3. Robinson (1965) independently proved the theorem of Marcienkiewicz—and was then told he had duplicated work published long before! Finally, Hänggi and Talkner (1980) emphasized that the theorem of Marcienkiewicz *implies* the theorem of Pawula.

We return to these theorems in the next chapter where we study the Fokker–Planck equation and the Kramers–Moyal expansion.

Cumulants: Combinations of the moments of a probability density function; i.e., the first cumulant is equal to the first moment, the second cumulant is related to the first two moments, the third cumulant is a function of the first three moments, etc. See van Kampen (1981) pp. 6–7.

Chapter 5

The Fokker–Planck Equation

The Fokker–Planck equation was originally developed as an alternative to the Langevin equation as a model of Brownian motion. Obukhov (1959) made the pioneering suggestion that the Fokker–Planck equation be used as a model of turbulent diffusion, but those who followed his suggestion from the 1960s into the 1980s chose to use the Langevin equation. This quotation from van Kampen (1992) p. 219 explains why—and gives a warning:

> "This approach is popular because it gives a more concrete picture than the Fokker–Planck equation, but it is mathematically equivalent to it. In nonlinear cases, it is subject to the same difficulties, and some additional ones."

This neglect of the Fokker–Planck equation ended when van Dop et al. (1985), Sawford (1986), and Thomson (1987) used it in their analyses of the Langevin equation as a model of turbulent diffusion. The authors of these and more recent papers have taken advantage of the different perspectives given by these two equations.

5.1 Mathematical properties

We begin with the companions, Eqs. (2.2) and (2.3), which we rewrite as follows:

$$du = (a_0 - a_1 u)dt + bdW(t) \qquad (5.1a)$$

and

$$\frac{\partial P(x,u,t|x_0,u_0,t_0)}{\partial t} + u\frac{\partial P(x,u,t|x_0,u_0,t_0)}{\partial x} =$$
$$-\frac{\partial}{\partial u}[(a_0 - a_1 u)P(x,u,t|x_0,u_0,t_0)]$$
$$+ \frac{1}{2}\frac{\partial^2}{\partial u^2}[b^2 P(x,u,t|x_0,u_0,t_0)].$$

$$(5.1b)$$

Here $P(x,u,t|x_0,u_0,t_0)$ is the joint *conditional* or *transition* probability density of the location x and velocity u at time t, given that it was at x_0 and u_0 at t_0. We learn in section 5.2 that there is a procedure for converting the joint *conditional* probability

density to the joint *unconditional* probability density $P(x,u,t)$ of the location x and velocity u at time t. According to van Kampen (1981) p. 209 and (1992) p. 193, the first-order differential term on the right has been called a "transport," "convection," or "drift" term, and the second-order differential term a "diffusion" or "fluctuation" term. In our application, with reference to Eqs. (3.1)–(3.2), we identify the symbol a_0 with the "drift correction," the quantity $a_1 u$ with the "fading memory" of the particle velocity (transport), and the coefficient $b^2/2$ with diffusion.

Sawford (1993) stated the following with respect to the Fokker–Planck equation as an Eulerian equivalent to the Lagrangian Langevin model of turbulent diffusion: 1) the distinction between Eulerian and Lagrangian statistics is merely one of sampling or conditioning, 2) the Fokker–Planck equation is equally valid for both statistics, 3) if the Eulerian statistics are known, they provide a constraint on the form of the equations (Fokker–Planck and Langevin), and 4) these equations are used to calculate Lagrangian statistics. We make the following comments to amplify these statements:

1) At a given location in space and instant in time, the Eulerian velocity is equal to the Lagrangian velocity of the particle at that point and time,

2) Equation (5.1b) may be written in terms of either the Eulerian $P_E(x,u,t)$ or the Lagrangian $P(x,u,t|x_0,u_0,t_0)$ with an appropriate choice of boundary conditions,

3) We use Lagrangian statistics to solve for the quantity $b^2/2$ in Eq. (5.1b) and Eulerian statistics to solve for a_0 and a_1, and

4) We use these solutions for a_0, a_1, and $b^2/2$ in the Langevin equation to calculate the Lagrangian trajectories of particles.

Thomson (1984) section 5 demonstrated that the coefficients a_0 and $b^2/2$ in Eq. (2.2) are respectively related to the first and second moments of the random forcing function in the Langevin equation.

He did this by using moment-generating functions to derive Eq. (3.2) as a solution for one-dimensional diffusion in inhomogeneous turbulence. Like Thomson (1987), we use the Fokker–Planck equation in section 7.2 to obtain the solutions for a_0 and a_1, and Kolmogorov's second hypothesis to obtain $b^2/2$.

5.2 Derivation

If we begin with fundamentals, we must consider three processes and their equations in deriving the Fokker–Planck equation:

1) The Markov process that is defined, in part, by the Chapman–Kolmogorov equation,

2) The jump process and its master equation, and

3) The diffusion process as modeled by the Fokker–Planck equation.

These processes and equations were presented by Wang and Uhlenbeck (1945) in sections 3, 5, and 8, respectively. Van Kampen (1981) presented material on processes 1–3 in sections IV.1–4, V.1, and VIII.1–2, respectively. Gardiner (1983) discussed processes 1–3 in sections 3.2.1, 3.5.1, and 3.5.2, respectively. The following is based largely on van Kampen because we found that text to be more useful for our purposes than the other references.

We begin by mathematically defining the process formulated by Markov in 1906 that is named after him. It is defined by two probability functions: the *probability density*,

$$P_1(y_1, t_1) \geq 0, \tag{5.2}$$

for the process to be in state y_1 at time t_1, and the *conditional* or *transition probability density*,

$$P_{1|1}(y_2, t_2 | y_1, t_1) \geq 0, \tag{5.3}$$

for the change from y_1 at t_1 to y_2 at t_2. The times in the Markov process are ordered as

$$t_1 < t_2 < t_3, \quad \text{etc.} \tag{5.4}$$

The probability functions Eqs. (5.2) and (5.3) cannot be arbitrary but *must follow* the following *two* constraints if the process is to be Markovian:

1) The probability density of y_2 at time t_2 is equal to the integral over all values of y_1 of the product of Eqs. (5.2) and (5.3):

$$P_1(y_2, t_2) =$$
$$\int P_{1|1}(y_2, t_2 | y_1, t_1) P_1(y_1, t_1) dy_1. \tag{5.5}$$

This, the first condition for a Markov process, states that the present state y_2 can have its origin in *any* of the probable states y_1 in the past. This condition *also* applies to non-Markovian processes.

2) A process starting at time t_1 from a value y_1 reaches a value y_3 at time t_3 by way of *any* of the possible values of y_2 at time t_2:

$$P_{1|1}(y_3, t_3 | y_1, t_1) =$$
$$\int P_{1|1}(y_3, t_3 | y_2, t_2) P_{1|1}(y_2, t_2 | y_1, t_1) dy_2. \tag{5.6}$$

This, the second condition, statistically decouples the process value y_3 at time t_3 (the future) from its value y_1 at time t_1 (the past) because, for a given value of y_1 at t_1, there is a wide range of possible values for y_2 at time t_2 (the present). This condition imposes the Markov assumption $P(y_3|y_2, y_1) = P(y_3|y_2)$ for $t_3 > t_2 > t_1$. The generalized extension of the preceding for the Markov process is

$$P(y_n|y_{n-1}, y_{n-2}, \ldots, y_2, y_1) = P(y_n|y_{n-1})$$

for $t_n > t_{n-1} > t_{n-2}, \ldots, t_3 > t_2 > t_1$. Equation (5.6), the *Chapman–Kolmogorov equation*, applies to any *Markov process*.

In summary for the Markov process: There is neither a unique past nor a unique future for a given present. In addition, the future is completely independent of the past but is a statistical function of the present.

The next step in the derivation of the Fokker–Planck equation involves Markov processes whose transition depends on the time difference alone, not on the successive times t_1 and t_2 (see sections IV.3–4 in van Kampen). From Eq. (5.3) for the transition probability density,

$$P_{1|1}(y_2, t_2 | y_1, t_1) = T_\tau(y_2 | y_1), \qquad (5.7)$$

where

$$\tau = t_2 - t_1. \qquad (5.8)$$

The Chapman–Kolmogorov equation, Eq. (5.6), becomes

$$T_{\tau+\tau'}(y_3 | y_1) = \int T_{\tau'}(y_3 | y_2) T_\tau(y_2 | y_1) dy_2 \qquad (5.9)$$

with

$$\tau, \tau' > 0. \qquad (5.10)$$

We now come to a difficult part of the derivation for the evolution of the Chapman–Kolmogorov equation for $T_{\tau+\tau'}$ as $\tau' \to 0$. Van Kampen gave the following balance equation for the transition probability $T_{\tau'}$ for small values of τ' [his Eq. (V.1.1)] which, in turn, is based on his Eq. (IV.6.8), a model of the process of radioactive decay:

$$T_{\tau'}(y_2 | y_1) =$$
$$(1 - a_0\tau')\delta(y_2 - y_1) + \tau'W(y_2, y_1) + O(\tau'),$$
$$(5.11)$$

where δ is the Dirac delta function and $O(\tau)/\tau' \to 0$ as $\tau' \to 0$. $W(y_2 | y_1)$ is the *transition probability per unit time* for the process from y_1 to y_2 with

$$W(y_2 | y_1) \geq 0. \qquad (5.12)$$

The quantity $1 - a_0\tau'$ is the probability that no transition takes place during the interval τ' so

$$a_0(y_1) = \int W(y_2 | y_1) dy_2. \qquad (5.13)$$

Similar material on the Poisson process, a Markov process that is homogeneous in time, was given by Feller (1968) in section XVII.2, and his Eq. (2.5) is equivalent to our Eq. (5.11). He explained that the transition from y_1 to y_2 can happen in three mutually exclusive ways. The first term on the right in Eq. (5.11) is the probability that the process is in state y_2 at the beginning of the interval τ' and stays there. The second term is the probability that the process jumps from y_1 to y_2 during τ'. The third term, $O(\tau')$, is the much smaller probability that the process will make more than one jump during τ' to get to y_2. Equation (5.11) for the *jump process* has been known for a long time because it is on p. 272 of the still-in-print classic by Doob (1953).

We substitute $(y_2 | y_1)$ for $(y_3 | y_2)$ in Eq. (5.11) and combine the result with Eq. (5.9) to get

$$T_{\tau+\tau'}(y_3 | y_1) =$$
$$\int [1 - a_0(y_3)\tau']\delta(y_3 - y_2)T_\tau(y_2 | y_1) dy_2$$
$$+ \int \tau'W(y_3 | y_2)T_\tau(y_2 | y_1) dy_2$$
$$+ \int O(\tau')T_\tau(y_2 | y_1) dy_2.$$
$$(5.14)$$

We evaluate the first integral with the Dirac delta function to get the result

$$T_{\tau+\tau'}(y_3 | y_1) =$$
$$[1 - a_0(y_3)\tau']T_\tau(y_3 | y_1)$$
$$+ \int \tau'W(y_3 | y_2)T_\tau(y_2 | y_1) dy_2$$
$$+ \int O(\tau')T_\tau(y_2 | y_1) dy_2.$$
$$(5.15)$$

We change from y_1 to y_3 in Eq. (5.13) and substitute the result for $a_0(y_3)$ into Eq. (5.15):

$$T_{\tau+\tau'}(y_3 | y_1) - T_\tau(y_3 | y_1) =$$
$$- \int \tau'W(y_2 | y_3)T(y_3 | y_1) dy_2$$
$$+ \int \tau'W(y_3 | y_2)T(y_2 | y_1) dy_2$$
$$+ \int O(\tau')T_\tau(y_2 | y_1) dy_2.$$
$$(5.16)$$

We divide by τ', apply the condition $O(\tau)/\tau' \to 0$ as $\tau' \to 0$ [see Eq. (5.11)] and get van Kampen's Eq. (V.1.4):

$$\frac{\partial}{\partial t}T_\tau(y_3 | y_2) =$$
$$\int [W(y_3 | y_2)T_\tau(y_2 | y_1) - W(y_2 | y_3)T_\tau(y_3 | y_1)] dy_2.$$
$$(5.17)$$

This is the *master equation*, a differential form of the *Chapman–Kolmogorov equation* for a *Markov process*. Van Kampen stated that it is usually written in a simplified, more intuitive form,

$$\frac{\partial P(y, t)}{\partial t} = \int [W(y | y')P(y', t) - W(y' | y)P(y, t)] dy',$$
$$(5.18)$$

which is his Eq. (V.1.5). He significantly revised the explanatory paragraph containing this equation for the 1992 edition. He stated that the solution of Eq. (5.18) is the transition probability $T_{t-t_1}(y|y')$ of the Markov process for *any choice* of t_1 and y_1. It is a gain-minus-loss equation for the probabilities of separate states y where the first term is the gain of state y as a result of transitions from other states y' and the second term is the loss from transition from state y to other states.

We finish the derivation of the Fokker–Planck equation by following section VIII.2 in van Kampen. We redefine the transition probability $W(y|y')$ in terms of the size r and origin y' of the jump:

$$W(y|y') = W(y'; r), \qquad (5.19)$$

where

$$r = y - y'. \qquad (5.20)$$

The master equation, Eq. (5.18), is then

$$\frac{\partial P(y,t)}{\partial t} =$$
$$\int W(y - r; r)P(y - r, t)dr - P(y, t)\int W(y; -r)dr.$$

$$(5.21)$$

Our first and basic assumption is that only small jumps occur so $W(y'; r)$ is a sharply peaked function of r but varies slowly with y':

$$W(y'; r) \approx 0 \quad \text{for} \quad |r| > \delta, \qquad (5.22a)$$

$$W(y' + \Delta y; r) \approx W(y'; r) \quad \text{for} \quad |r| < \delta. \qquad (5.22b)$$

Our second assumption is that the solution $P(y, t)$ also varies slowly with y'. Then it is possible to *approximate* the product $W(y - r; r)P(r - r, t)$ in the first integral in Eq. (5.21) by means of a Taylor series from the zeroth through the second order:

$$W(y - r; r)P(y - r, t) =$$
$$W(y; r)P(y, t) - r\frac{\partial}{\partial y}[W(y; r)P(y, t)]$$
$$+ \frac{r^2}{2}\frac{\partial^2}{\partial y^2}[W(y; r)P(y, t)].$$

$$(5.23)$$

We combine Eqs. (5.21) and (5.23) to get

$$\frac{\partial P(y,t)}{\partial t} = \int W(y; r)P(y, t)dr$$
$$- \int r\frac{\partial}{\partial y}[W(y; r)P(y, t)]dr$$
$$+ \frac{1}{2}\int r^2\frac{\partial^2}{\partial y^2}[W(y; r)P(y, t)]dr$$
$$- P(y, t)\int W(y; -r)dr.$$

$$(5.24)$$

The first and last terms on the right cancel. In analogy to Eq. (5.13), we introduce the moments

$$\mu_n(y) = \int_{-\infty}^{\infty} r^n W(y; r)dr. \qquad (5.25)$$

Then we get the *Fokker–Planck equation*, an *approximation* to the master equation, Eq. (5.18)—unless P is Gaussian in which case it is exact:

$$\frac{\partial P(y, t|y_1, t_1)}{\partial t} = -\frac{\partial}{\partial y}[\mu_1(y)P(y, t|y_1, t_1)] +$$
$$\frac{1}{2}\frac{\partial^2}{\partial y^2}[\mu_2(y)P(y, t|y_1, t_1)],$$

$$(5.26)$$

which is equivalent to Eq. (VIII.2.5) in van Kampen. The solution of Eq. (5.26) is for the *transition probability density* $P(y, t|y_1, t_1)$ for $t \geq t_1$ that at t_1 reduces to $\delta(y - y_1)$ (van Kampen 1981, p. 210; 1992, p. 193). Furthermore, $P(y, t|y_1, t_1)$ is also the *conditional* probability density that can be replaced with the *unconditional* probability density $P(y, t)$ by means of Eq. (5.5). It is not always clear which usage, *conditional* or *unconditional*, is implicit in the many references that present the Fokker–Planck equation only in terms of $P(y, t)$. These distinctions are made for the conditional Lagrangian P_L and unconditional Eulerian P_E in Borgas and Sawford (1994a).

Van Kampen stated in section VIII.1 that this equation is an attractive substitute for the master equation because the moments $\mu_1(y)$ and $\mu_2(y)$ in Eq. (5.26) are much easier to determine than the kernel $W(y|y')$ in Eq. (5.18). He added that the Fokker–Planck equation is an approximate substitute for the master equation, Eq. (5.17), for any Markov process in which the jumps [see Eq. (5.20)] are small. Fokker and Planck assumed small jumps. Kolmogorov (1931) assumed *infinitely small jumps* in his rigorous mathematical derivation. In recognition of his work, Eq. (5.26) is sometimes called the Fokker–Planck–Kolmogorov equation.

With reference to our discussion of Eq. (5.1), we identify the first moment $\mu_1(y)$ in Eq. (5.26) as a "drift coefficient" and the second moment $\mu_2(y)$ as a "diffusion coefficient." The *Fokker–Planck equation* is a model of a *diffusion* process in phase space.

5.3 The forward and backward equations

For the record, we identify and define the forward and backward equations (Kolmogorov 1931). The following is based largely on the first few pages of section VI.3, "The diffusion equations and the corresponding Markov processes," in Doob (1953)—see his Eqs. (3.1), (3.4), and (3.4′). [Similar material was presented by Thomson (1987) sections 2.2 and 2.3.] The first equation is the familiar Langevin equation,

$$dx(t) = m[t,x(t)]dt + \sigma[t,x(t)]dW(t), \quad (5.27)$$

where the continuous Markov process $x(t_2) - x(t_1)$ is the sum of small increments $dx(t)$, each of which is Gaussian with mean mdt and variance $\sigma^2 dt$, and $dW(t)$ is the incremental Wiener process. The symbols m and σ have the same meaning in the following two equations.

Kolmogorov's backward equation is

$$\frac{\partial p(s,\xi;t,\eta)}{\partial t} = -m(s,\xi)\frac{\partial p}{\partial \xi} - \frac{\sigma(s,\xi)^2}{2}\frac{\partial^2 p}{\partial \xi^2}, \quad (5.28)$$

where $p(s,\xi;t,\eta)$ is a transition probability function. This is a parabolic partial differential equation in s and ξ for $s \leq t$. It is called the backward equation because the derivatives are taken with respect to the backward (earlier in time) variables.

His forward equation is identical to the Fokker–Planck equation,

$$\frac{\partial p'(s,\xi;t,\eta)}{\partial t} = -\frac{\partial}{\partial \eta}[m(t,\eta)p']$$
$$+\frac{1}{2}\frac{\partial^2}{\partial \eta^2}[\sigma(t,\eta)^2 p'] \quad \text{with} \quad p' = \partial p/\partial \eta.$$
$$(5.29)$$

This is a parabolic partial differential equation in t and η for $s \leq t$. It is called the forward equation because the derivatives are taken with respect to the forward (later in time) variables. Doob stated that

it is usually the more natural for use as a model of physical processes. We apply it to turbulent diffusion.

5.4 The Kramers–Moyal expansion

In the third and final stage of our derivation of the Fokker–Planck equation in Eqs. (5.19)–(5.26), we approximated the master equation, Eq. (5.18),

$$\frac{\partial P(y,t)}{\partial t} = \int [W(y|y')P(y',t) - W(y'|y)P(y,t)]dy', \quad (5.30)$$

with a Taylor series expansion truncated at the second moment to get the Fokker–Planck equation, Eq. (5.26):

$$\frac{\partial P(y,t)}{\partial t} =$$
$$-\frac{\partial}{\partial y}[\mu_1(y)P(y,t)] + \frac{1}{2}\frac{\partial^2}{\partial y^2}[\mu_2(y)P(y,t)].$$
$$(5.31)$$

The Kramers–Moyal expansion (van Kampen 1981, section VIII.2; Gardiner 1983, section 7.2.2) continues the Taylor series beyond the second moment:

$$\frac{\partial P(y,t)}{\partial t} = \sum_{n=1}^{\infty} \frac{(-1)^n}{n!}\frac{\partial^n}{\partial y^n}[\mu_n(y)P(y,t)]. \quad (5.32)$$

This is formally the same as Eq. (5.18), the master equation. If the random forcing in the companion stochastic differential equation (i.e., the Langevin equation) is Gaussian, the Kramers–Moyal expansion truncates to the Fokker–Planck equation and $P(y,t)$ is Gaussian. If the random forcing is not Gaussian, $\mu_n \neq 0$ for $n \geq 3$ and $P(y,t)$ is not Gaussian. It is appropriate to discuss some implications of Eq. (5.31) versus Eq. (5.32) with respect to both the input (stochastic forcing) and the output (velocity fluctuations) of the Langevin equation as a model of turbulent diffusion.

The Taylor series in Eq. (5.32) has, in general, an infinite number of nonzero terms. However, in practice it is necessary to truncate this equation after a finite number of moments (say three or four) because of lack of knowledge of the higher moments. As we noted in section 4.3, a probability density defined by Eq. (5.32) that is truncated after the third moment or higher is not realizable (Marcienkiewicz 1938; Pawula 1967a,b). Only the complete series with $n \to \infty$ in Eq. (5.32) gives a realizable probability distribution function if the distribution is not Gaussian. In contrast, the Gaussian distribution is *completely* determined by its first two moments. Gardiner and van Kampen discussed the nonrealizability of third and higher order truncations of Eq. (5.49) in view of the proofs of Marcienkiewicz and Pawula, and agreed that such truncations could be useful—if applied with caution.

Thomson (1987) stated that it may be possible to use non-Gaussian forcing (in the Langevin equation) to obtain satisfactory results, but the criteria for a properly designed model (the well-mixed condition, a continuous Markov process for the output) are satisfied only approximately. This is illustrated for the well-mixed condition in section 3 of Thomson (1984) by runs 3 and 4 that are plotted in Fig. 2 in that paper. The mathematically required non-Gaussian forcing ($\mu_n \neq 0$ for $n \leq 3$ and $\mu_n = 0$ for $n \geq 4$) was only *approximated* in those numerical simulations. Sawford and Borgas (1994) demonstrated that the stochastic velocity process is continuous (and consistent with Kolmogorov's hypotheses) *only* with Gaussian forcing. They used a differential form of the master equation [the differential Chapman–Kolmogorov equation, Eq. (3.4.22) in Gardiner (1983)] that treats separately the continuous and discontinuous parts of the stochastic process. With reference to Eqs. (5.30)–(5.32), this differential equation may be written as follows:

$$\frac{\partial P(y,t)}{\partial t} =$$
$$-\frac{\partial}{\partial y}[\mu_1(y)P(y,t)] + \frac{1}{2}\frac{\partial^2}{\partial y^2}[\mu_2(y)P(y,t)]$$
$$+ \int [W(y|y')P(y',t) - W(y'|y)P(y,t)]dy'.$$

$$(5.33)$$

The first two lines in Eq. (5.33) are the Fokker–Planck equation for continuous processes, and the third line accounts for discontinuities. Here $\mu_1(y)$ represents "drift" and $\mu_2(y)$ "diffusion." $W(y|y')$ and $W(y'|y)$ are transition probabilities for *finite* changes in y. Equation (5.33) is formally equivalent to the Kramers–Moyal expansion, Eq. (5.32). We may then infer that the higher-moment terms with $n \geq 3$ in Eq. (5.32) correspond to the second line for discontinuous processes in Eq. (5.33). We now have a complete explanation of Thomson's (1987) comment: "Such a model must either be non-existent (in the sense that no random forcing exists with the required moments) or have discontinuous phase-space trajectories."

We restrict ourselves to Gaussian distributions in all the following, except for chapter 10. In that chapter, we learn how the Fokker–Planck equation can be used to obtain a Langevin equation with Gaussian forcing and non-Gaussian output.

Chapter 6

Stochastic Differential Equations for Turbulent Diffusion

6.1 Langevin equation for stationary, inhomogeneous, Gaussian turbulence

We begin with the model proposed by Wilson et al. (1983) and rigorously derived by Thomson (1984, 1987):

$$dw = -\frac{w}{\tau_L}dt$$
$$+ \frac{1}{2}\left[1 + \left(\frac{w}{\sigma_w}\right)^2\right]\frac{\partial \sigma_w^2}{\partial z}dt + \left(\frac{2\sigma_w^2}{\tau_L}\right)^{1/2} dW(t).$$

$$(6.1a)$$

Its companion is

$$dz = wdt. \qquad (6.1b)$$

As we demonstrate in section 7.2, this stochastic Lagrangian model for one-dimensional (vertical) diffusion in stationary inhomogeneous turbulence meets Thomson's (1987) well-mixed design criterion, and both its input (random forcing) and its output (velocity fluctuations) are Gaussian. We note that Pope (1987) commented that 1) experimental measurements demonstrate that inhomogeneous turbulence is non-Gaussian and 2) theory indicates that inhomogeneous turbulence that is initially Gaussian does not stay Gaussian. However, we defend this model as an extremely useful fiction or idealization.

We learned in chapter 3 that the physical basis of this equation is Kolmogorov's (1941) hypotheses for the local properties of turbulence at very large Reynolds numbers. Let us review the physical interpretation of the terms on the right in Eq. (6.1a). The first two terms are deterministic and the third is stochastic. The first term represents the "fading memory" of the particle velocity. The second term is the "drift correction" for inhomogeneous turbulence; without it particles that were initially well-mixed would concentrate in regions of low velocity variance. The third (stochastic) term is a model of random velocity fluctuations with short correlation times on the order of the Kolmogorov timescale.

From chapter 4, we know that Eq. (6.1a) is a stochastic differential equation that, because of

the properties of the incremental Wiener process $dW(t)$, models a *diffusion* process that is *continuous* (but not differentiable), *Gaussian*, and *Markovian*.

If the Lagrangian timescale τ_L and the velocity variance σ_w^2 are functions of the vertical coordinate z, Eq. (6.1a) is a *nonlinear* equation for the velocity increment $dw(z,t)$. As van Kampen bluntly noted on p. 244 of the 1981 edition, "However, [Eq. (6.1a)] *as it stands has no meaning* (emphasis in original)." He was not quite as emphatic in sections IX.4–5 of the 1992 edition where he discussed at more length the following question: What values of $[\tau_L$ and $\sigma_w^2]$ should be used in integrating this equation? With $dt = t_2 - t_1$, should it be at time t_1 (and z_1) or at time t_2 (and z_2) or a time halfway between (and at a corresponding value of z)? We answer this question in the next section.

6.2 The Ito and Stratonovich interpretations

More than forty years ago (and about forty years after Langevin published his paper), Ito published several papers in Japan and one in the United States (Ito 1951), that put stochastic differential equations on a rigorous foundation. The results of Ito's work were incorporated in section VI.3 of Doob (1953). We begin with Eq. (VI.3.1) on p. 273 in Doob, rewritten with our symbol for the incremental Wiener process [Doob's $dy(t) \equiv dW(t)$]:

$$dx(t) = m[t, x(t)]dt + \sigma[t, x(t)]dW(t), \qquad (6.2)$$

where, as we noted for Eq. (5.27), the continuous Markov process $x(t_2) - x(t_1)$ is the sum of small increments $dx(t)$, each of which is Gaussian with mean mdt and variance $\sigma^2 dt$. [Here $x(t)$ is a general stochastic variable, not position.] The proper interpretation of Eq. (6.2) is illustrated by Doob's Eq. (VI.3.1′) on p. 277:

$$x(t) - x(a) =$$
$$\int_a^t m[s, x(s)]ds + \int_a^t \sigma[s, x(s)]dW(s),$$

$$(6.3)$$

where the range of t is in the finite interval $[a, b]$: $a < t < b$. That is, Eq. (6.2) should be regarded as mathematical shorthand for Eq. (6.3), with the small increment $x(t) - x(a)$ replacing $dx(t)$. There is no problem in the evaluation of the first term on the right; the classical Riemann–Stieltjes calculus is applicable. But how is the second integral to be interpreted? What value of $\sigma[s, x(s)]$ in the interval $a < s < t$ should be used? Ito chose the following [from Eq. (VI.3.6) on p. 279]:

$$x_n(t) = x(a) + \int_a^t m[s, x_{n-1}(s)]ds$$

$$+ \int_a^t \sigma[s, x_{n-1}(s)]dW(s),$$

$$(6.4)$$

that is, the *initial* value of $\sigma[s, x(s)]$. Therefore the Ito integral is a *nonanticipating* function.

This work of Ito and Doob did not settle the matter. In general, during the following decade, mathematicians applied the above Ito calculus while physicists and engineers used classical Riemann–Stieltjes calculus with a second integral redefined to be symmetric in time (Gray and Caughey 1965):

$$x_n(t) = x(a) + \int_a^t m[s, x_n(s)]ds$$

$$+ \int_a^t \sigma[s, \frac{1}{2}[x_{n-1}(s) + x_n(s)]]dW(s).$$

$$(6.5)$$

That is, the value of $\sigma[s, x(s)]$ at the *midpoint* of the interval $a < s < t$ is used. In contrast to the Ito integral above, the Stratonovich integral is an anticipating function. The Fokker–Planck equation corresponding to Eq. (6.4) has one form and that corresponding to Eq. (6.5) another. Consequently, different sets of results were obtained for apparently identical problems. This was because the differences between a strictly defined mathematical problem and a mathematical approximation to a physical problem were not always understood. Gray and Caughey concluded that "apparently identical" problems are not necessarily identical, that there are instances when the mathematical (Ito) approach is appropriate, and that the physical approach should be used when the Wiener process is not a good approximation to the random excitation. We find later in this section that problems can be formulated to obtain the same result with either integral.

Gray and Caughey were apparently unaware that Stratonovich had proposed the use of Eq. (6.5) in "A new representation for stochastic integrals and equations." This paper was published in Russian in 1964 and in English (Stratonovich 1966; see also Stratonovich 1968, sections 2.1–2.2). His motive was to replace the Ito calculus with a method that used the familiar differential and integral calculus. He developed his approach as a result of working with smoothed (not completely Markovian) processes and conditional Markov processes. He gave a solid mathematical foundation to what has become known as the Stratonovich calculus, and he showed how it is related to the earlier work of Ito and Doob. His definition of the stochastic integral in Eq. (6.5), written in (approximately) his nomenclature, is

$$\int_a^t \sigma\left[s_{n-1}, \frac{1}{2}[x_{n-1}(s) + x_n(s)]\right] dW(s) =$$

$$\lim_{\Delta \to 0} \sum_{n=2}^N \sigma\left[s_{n-1}, \frac{1}{2}[x_{n-1}(s) + x_n(s)]\right]$$

$$\times [W(s_n) - W(s_{n-1})],$$

$$(6.6)$$

where

$$a = s_1 < s_2 < \ldots < s_N = b$$

$$\text{and} \quad \Delta = \max(s_n - s_{n-1}).$$

$$(6.7)$$

Note that the symbols for the incremental Wiener process on each side of Eq. (6.6) are consistent with the identity in Eq. (4.3b). Equation (6.7) is not only applicable to Eq. (6.6), but also to Eqs. (6.3)–(6.5). The sequence of times s_n in the finite interval $[a, b]$ is closely packed, and in the limit $N \to \infty$ as $\Delta \to 0$. Similarly Stratonovich defined the Ito form of the stochastic integral in Eq. (6.4) as

$$\int_a^t \sigma[s_{n-1}, x_{n-1}(s)]dW(s) =$$

$$\lim_{\Delta \to 0} \sum_{n=2}^N \sigma[s_{n-1}, x_{n-1}(s)][W(s_n) - W(s_{n-1})].$$

$$(6.8)$$

Gardiner (1990) p. 101 stated that the more "obvious" definition of $\sigma[s, x(s)]$ in terms of the midpoint of time $s = (s_{n-1} + s_n)/2$ was not used by Stratonovich, although that definition is widespread (for example, Øksendal 1992, p. 17). Stratonovich

(1966, 1968) considered both but stated that the useage in Eq. (6.6) is possible if $\sigma[s, x(s)]$ is differentiable in s (see also Gardiner 1990, section 4.3.6).

Let us create a "word picture" of the summations in Eqs. (6.6) and (6.8). In section 4.2, we described white noise as a densely packed sequence of positive and negative impulses, and we defined the Wiener process as the continuous *but not differentiable* integral of white noise. We may visualize the incremental process $[W(s_n) - W(s_{n-1})]$ as a random, irregular, jagged path with each abrupt, *discontinuous* turn at the times s_{n-1}, s_n, s_{n+1}, etc., as illustrated in Fig. 4.3. These adjectives for the path and its turns apply at all scales from the order of the interval $[a, b]$ down to the infinitesimal. The process defined by Eq. (6.6) is not identical to that of Eq. (6.8). One reason is the difference between $\sigma[s_{n-1}, [x_{n-1}(s) + x_n]/2]$ and $\sigma[s_{n-1}, x_{n-1}(s)]$. Another is that the incremental Wiener process $dW(t)$ is independent of $W(t)$ in the Ito calculus but is not in the Stratonovich interpretation (Mortensen 1969, p. 285; Gardiner 1990, section 4.2.3).

Stratonovich gave the following to relate his integral to that of Ito (in our notation):

$$(I) \int_a^t \sigma[s_{n-1}, x_{n-1}(s)]dW(s) =$$

$$(S) \int_a^t \sigma\left[s_{n-1}, \frac{1}{2}[x_{n-1}(s) + x_n(s)]\right] dW(s)$$

$$-\frac{1}{2}\int_a^t \sigma[s_{n-1}, x_{n-1}(s)]\frac{\partial}{\partial x}\sigma[s_{n-1}, x_{n-1}(s)]ds.$$

$$(6.9)$$

Here (I) indicates the Ito stochastic integral and (S) the Stratonovich stochastic integral. Equivalents of this equation are given in Arnold (1974) section 10.2 and Gardiner (1983, 1990) section 4.3.6, and are given implicitly in van Kampen (1992) section IX.4 and Øksendal (1992) chapter III. The second term on the right is an ordinary integral and a drift term; van Kampen (1992) pp. 232–233 called it a "spurious drift." Equation (6.9) gives us a practical reason for using the Ito integral instead of that of Stratonovich: If we use the latter, we would have to add a second drift correction to the one included in Eq. (6.1a). Arnold listed (pp. 167–168) the advantages and disadvantages of each integral. The Ito integral has these advantages:

1)   There are simple formulas for the first two moments of the integral.

2)   A stochastic differential equation based on this integral yields a diffusion process as a solution.

3) This process is a martingale.

Its disadvantage is that the rules for differentiation and integration differ from those of the familiar calculus. The Stratonovich calculus follows the familiar rules but has none of the above advantages. Mortensen (1969) stated in section 6 that Stratonovich did not define a new stochastic integral; rather the new integral was defined in terms of the Ito integral. Mortensen also stated that the Ito integral is more fundamental, more general, and more versatile than the Stratonovich integral, but that the Stratonovich integral is sufficiently versatile to be useful in the integration of stochastic differential equations. Malliaris (1983) noted on p. 494 that Ito's calculus is an independent, self-contained mathematical system, and that its nonanticipating property is appropriate for modeling economic uncertainty (the subject of that paper).

We have not defined the word "martingale" that appears in the above list of the advantages of the Ito integral. "Martingale" does not appear in the indexes of Gardiner (1983, 1990) or van Kampen (1981, 1992), but it does in Arnold (1974), Doob (1953), and Øksendal (1992). Øksendal (pp. 22–24) used the martingale property to prove that the Ito integral can be chosen to be a continuous function of time. Doob (section VII.2) and Arnold (section 1.9) stated that martingales are a mathematical abstraction of a "fair game" in gambling. Martingales are discussed in this context in Feller (1971) p. 210, but the example on p. 399 in Feller (1968) of Markov chains that are also martingales is more pertinent to diffusion. Let us assume a finite chain with states numbered $E_0, E_1, \ldots, E_i, \ldots, E_a$, and that the states E_0 and E_a are absorbing, not reflecting. Then all the interior states $E_1, \ldots, E_{a-1}$ are transient and the process will eventually terminate at either E_0 or E_a. If the process starts with the interior state E_i, the probabilities of absorption at E_0 or E_a are $1 - i/a$ and i/a, respectively. If the initial state is closer to one of the two absorbing boundaries, the probability is greater that the process will end there. In other words, there is no preferential "drift" one way or the other. In this sense, the "marked particles" diffusing in our turbulent flow are in a "fair game" that is controlled by the Ito integral.

There is general agreement among Gray and Caughey (1965), Mortensen (1969), Arnold (1974), van Kampen (1981, 1992), Gardiner (1983, 1990), and Øksendal (1992) that there are valid applications for both integrals, Ito and Stratonovich, and that one is more suitable for some applications than the other. We note that Eq. (6.9) can be used to convert an Ito integral to a Stratonovich integral (and vice versa), so *either* method can be used to obtain the same result. We follow van Dop et al. (1985), Sawford (1986), and Thomson (1987) who specifically mentioned and used the Ito calculus in

their work toward a rigorous foundation for stochastic Lagrangian models of diffusion in turbulence.

Two statements of the obvious: First, there is no difference between the Stratonovich and Ito integrals if the coefficient of the incremental Wiener process is constant. Second, discrete numerical calculations with the increments Δt and $\Delta W(t)$ are simpler with the Ito integral.

Ito integral: A stochastic integral in which the value of the variable coefficient $\sigma[s, x(s)]$ of the incremental Wiener process $dW(s)$ is taken at the *beginning* of the transition from the state $x_{n-1}(s)$ to $x_n(s)$. It is therefore a nonanticipating function in that it is "ignorant" of its next state $x_n(s)$. It is a part of the Ito calculus that does not follow the rules of the ordinary calculus.

Stratonovich integral: A stochastic integral in which the value of the variable coefficient $\sigma[s, x(s)]$ is taken at the *midpoint* of the transition from $x_{n-1}(s)$ to $x_n(s)$. Therefore, it is an anticipating function in that it is "aware" of its next state. The Stratonovich integral follows the rules of the ordinary calculus.

Martingale: A stochastic process with a set of possible states E_{n-1}, E_n, E_{n+1}, etc. The transition from state E_n to state E_{n-1} is just as likely as that from E_n to E_{n+1}. The Ito integral is a martingale, but the Stratonovich integral is not because of the spurious drift term [see Eq. (6.9)].

6.3 The diffusion limit:
The random displacement equation

For convenience, we repeat Eqs. (6.1a,b), the LEM for stationary inhomogeneous turbulence originally proposed by Wilson et al. (1983) and rigorously derived by Thomson (1984, 1987):

$$dw = -\frac{w}{\tau_L} dt +$$
$$\frac{1}{2}\left[1 + \left(\frac{w}{\sigma_w}\right)^2\right]\frac{\partial \sigma_w^2}{\partial z} dt + \left(\frac{2\sigma_w^2}{\tau_L}\right)^{1/2} dW(t),$$

$$(6.10a)$$

and

$$dz = w\,dt. \qquad (6.10b)$$

These equations are used in two steps to calculate the random displacements of marked particles in turbulent flow. First, Eq. (6.10a) is integrated to obtain a solution for the velocity $w(z, t)$, and then this value of $w(z, t)$ is used to integrate Eq. (6.10b) to obtain the displacement $z(t)$. We now consider the random displacement model in which the location $z(t)$ is calculated directly. Under the condition that the timescale $\tau_L \to 0$ (or $t/\tau_L \to \infty$), the two above equations can be combined and mathematically transformed into a stochastic differential equation for $z(t)$. This is called the Markov or diffusion equation limit of the Langevin equation model.

Arnold (1974) pp. 134–136 converted the Gaussian process for $w(t)$ into a Gaussian process for $z(t)$. Boughton et al. (1987) presented essentially the same mathematics in their proposal that the random displacement model be applied to turbulent diffusion in the atmosphere. These cases considered the equivalent of Eqs. (6.10a,b), but without the drift correction for inhomogeneous turbulence [the middle term on the right in Eq. (6.10a)]. Transformations have also been published for stationary inhomogeneous turbulence. Durbin (1984) transformed Eqs. (6.10a,b) into

$$dz = \frac{\partial}{\partial z}(\sigma_w^2 \tau_L) dt + (2\sigma_w^2 \tau_L)^{1/2} dW(t). \qquad (6.11)$$

He stated that one of the constraints on a consistently formulated Langevin model is that it must reduce to an eddy diffusivity model [the Eulerian equivalent of Eq. (6.11) as $\tau_L \to 0$ or $t/\tau_L \to \infty$]. Van Dop et al. (1985) presented in their Eq. (48) the Markov limit of a Langevin model that included the drift correction of Legg and Raupach (1982). Their Eq. (48) can be converted into our Eq. (6.11) exactly, yet the Legg and Raupach model does not meet Thomson's "well-mixed" criterion. Thomson demonstrated in section 3.5 of his 1987 paper that

1) A Langevin equation model that meets the well-mixed condition reduces to a diffusion equation as $\tau_L \to 0$, and

2) A Langevin model that reduces to a diffusion equation *does not necessarily* meet the well-mixed condition.

De Baas and Troen (1989) showed that the Wilson et al. (1983) model [Eq. (6.10a,b)] and the Legg and Raupach model reduce to exactly the same Markov limit. In the following, we outline the procedure they used (but only for the Wilson et al. model), and mark the step in the derivation at which the difference between the two models disappears.

Equation (6.10a) can be rewritten as

$$d\left(\frac{w}{\sigma_w}\right) = \left(-\frac{w}{\sigma_w \tau_L} + \frac{\partial \sigma_w}{\partial z}\right) dt + \left(\frac{2}{\tau_L}\right)^{1/2} dW(t).$$

$$(6.12)$$

We use the scaling relations

$$\tau_L = \alpha T \quad \text{and} \quad \sigma_w = \beta \Sigma, \qquad (6.13)$$

where T and Σ (as well as τ_L and σ_w) are functions of z and integrate the result to get

$$\alpha\left[\left(\frac{w}{\Sigma}\right)_t - \left(\frac{w}{\Sigma}\right)_0\right] =$$
$$\int_0^t \left[\left(-\frac{w}{\Sigma T} + \alpha\beta^2 \frac{d\Sigma}{dz}\right) dt + \left(\frac{2\alpha\beta^2}{T}\right)^{1/2} dW(t)\right].$$

$$(6.14)$$

Arnold (1974), Boughton et al. (1987), and de Bass and Troen (1989) required the following or its equivalent:

$$\alpha \to 0, \ \beta \to \infty \quad \text{such that}$$
$$\alpha\beta^2 = K, \quad \text{a constant}.$$

$$(6.15)$$

The result is that σ_w increases less rapidly than τ_L decreases as the turbulence field is rescaled. De Baas and Troen stated that these constraints are required in order that the limiting conditions of $\tau_L \to 0$ with t fixed and $t/\tau_L \to \infty$ with τ_L fixed be equivalent. The term on the left in Eq. (6.14) is, under these conditions, bounded in the "mean square" sense for all time (and approaches zero).

We combine Eqs. (6.10b), (6.14), and (6.15) with the result that

$$\int_{z_0}^{z_1} \frac{dz}{\Sigma T} = \int_0^t \left[K \frac{\partial \Sigma}{\partial z} dt + \left(\frac{2K}{T}\right)^{1/2} dW(t)\right].$$

$$(6.16)$$

We convert Eq. (6.16) to a stochastic differential equation by differentiating both sides. We use the rules of the Ito calculus (Arnold, pp. 89–92) to differentiate the left side. This involves the use of the first two terms of the Taylor series,

$$f(x) - f(a) =$$
$$(x-a)f'(a) + \frac{(x-a)^2}{2!} f''(a) + \ldots,$$

$$(6.17)$$

where $(x - a)$ corresponds to dz and $f'(a)$ corresponds to $(\Sigma T)^{-1}$. The differential of Eq. (6.16) is therefore

$$\frac{dz}{\Sigma T} - \frac{1}{2}\left(\frac{dz}{\Sigma T}\right)^2 \frac{d}{dz}(\Sigma T) =$$
$$K \frac{d\Sigma}{ds} dt + \left(\frac{2K}{T}\right)^{1/2} dW(t).$$

$$(6.18)$$

This is of the quadratic form

$$A(dz)^2 + B(dz) + C = 0. \qquad (6.19)$$

In preparation for the next two equations, we set $B = -1$ and select the root

$$dz = [1 - (1 - 4AC)^{1/2}]/2A. \qquad (6.20)$$

We expand $(1 - 4AC)^{1/2}$ in the infinite series

$$(1 - 4AC)^{1/2} = 1 - \frac{4AC}{2} - \frac{(4AC)^2}{8} - \ldots, \ (6.21)$$

and retain only the first three terms. It is in the truncation at the third term that the *difference* between the models of Wilson et al. (1983) and Legg and Raupach (1982) *vanishes*. We retain the first three terms in accord with the rules of the Ito calculus that terms involving $\langle dW(t)\rangle^2$ be retained because

$$(dt)^2 = 0,$$

$$\langle dW(t)dt \rangle = 0, \qquad (6.22)$$

$$\langle dW(t) \rangle^2 = dt.$$

The final result of applying Eqs. (6.19)–(6.22) to Eq. (6.18) is

$$dz = \frac{\partial}{\partial z}(\sigma_w^2 \tau)dt + (2\sigma_w^2 \tau)^{1/2}dW(t), \qquad (6.23)$$

which is identical to Eq. (6.11). The first term on the right is the "drift correction."

6.4 Applicability of the Langevin and random displacement equations

Of the two models, one based on the Langevin equation and the other on random displacements, the former is more generally applicable because it can be used to model diffusion in both Gaussian and non-Gaussian turbulence for all times much greater than the Kolmogorov timescale τ_K,

$$t \gg \tau_K, \qquad (6.24)$$

provided that the turbulence is three-dimensional and at very high Reynolds number (see section 3.2). As a consequence of the limiting conditions imposed in the transformation from Eq. (6.12) to Eq. (6.23) ($\tau \to 0$ with t fixed and $t \to \infty$ with τ fixed), the times for which the random displacement model is applicable are much greater than the local Lagrangian velocity decorrelation timescale,

$$t \gg \tau_L. \qquad (6.25)$$

The classical eddy diffusivity (or K-theory) model can be used in violation of Eq. (6.25) by using (nonphysical) time-dependent coefficients $K(t)$. As we discuss further in chapter 10, the random displacement and eddy diffusivity models have limited applicability in the non-Gaussian convective boundary layer.

Chapter 7

Criteria for Stochastic Models of Turbulent Diffusion

7.1 Descriptions of criteria

We have mentioned two criteria for stochastic Lagrangian models of turbulent diffusion: Durbin's (1984) requirement that a Langevin model reduce to the diffusion equation and Thomson's (1987) well-mixed criterion. It is appropriate at this point to review the several criteria that have been proposed. These were the subject of Thomson's classic paper in which he considered the following five:

1) The well-mixed condition: If the particles of a tracer are initially well-mixed (in both position and velocity space) in a turbulent flow, will they remain so?

2) Is the small-time behavior of the velocity distribution of the particles from a point source correct?

3) Are the Eulerian equations derived from the Lagrangian model compatible with the true Eulerian equations?

4) Are the forward and reverse formulations of the dispersion consistent?

5) Does the model reduce to a diffusion equation as the Lagrangian timescale tends toward zero?

Thomson showed in section 3 of his paper that a generalized version of criterion 1 is sufficient to ensure that criteria 2–5 are satisfied. Specifically, he proved mathematically that criteria 2–4 are equivalent to criterion 1 and that criterion 5 is a weaker condition than criterion 1. [We stated in section 6.3 that a model that satisfies criterion 5 does not necessarily satisfy criterion 1.] In the following chapter, we use material from Thomson's sections 2–4 to expand upon his section 5 on the "simplest" solution for a Langevin model of nonstationary, inhomogeneous three-dimensional diffusion in Gaussian turbulence.

Based on the number of citations in the literature, Thomson's paper appears to have overshadowed a paper on the same subject by Pope (1987). The two papers addressed some of the same questions of mathematical consistency, but there are significant differences. Thomson based his Langevin model on an assumed generalized flow and its turbulence statistics. He did not specify the means of generating that flow. In contrast, Pope assumed a closed set of equations that, in principle, determines the properties of the flow and its turbulence statistics. As stated in Haworth and Pope (1986), the Navier–Stokes equation is taken as the starting point for this modeling, not Brownian motion. When allowance is made for these differences, the criteria of Pope and Thomson are not inconsistent.

7.2 Applying the well-mixed criterion

In the interest of simplicity, we consider one-dimensional (vertical) diffusion in constant-density flow. We assume that the vertical velocity fluctuations of the fluid and its population of marked particles (tracer) have the same statistical distribution, and that the unconditional density function of the tracer, $P_t(z, w, t)$, is a fraction of that of the fluid, $P_a(z, w, t)$ [see Thomson (1987) section 2.1]:

$$C(z, w, t) = P_t(z, w, t)/P_a(z, w, t), \qquad (7.1a)$$

where $C(z, w, t)$ is the concentration of the tracer in phase space. If the tracer is well-mixed,

$$C(z, w, t) = \text{constant}. \qquad (7.1b)$$

That is, the distribution of the tracer in z, w space is a constant, time-independent fraction of the distribution of the fluid in the same phase space. This means that once the tracer is well mixed, it stays that way. Then the Fokker–Planck equation for vertical diffusion may be written in terms of the density function of either the fluid or the tracer; the difference is a proportionality constant which need not be included. In the above and in Thomson, $P_t(z, w, t)$ and $P_a(z, w, t)$ are not strictly *probability* density functions in phase space. (Our definitions are slightly different from those in Thomson.) We return to "standard" probability density functions in the following.

We begin with the Langevin equation for one-dimensional (vertical) diffusion in nonstationary, inhomogeneous turbulence,

$$dw = a(z, w, t)dt + b(z, w, t)dW(t), \qquad (7.2a)$$

where dw is a random vertical Lagrangian velocity increment, $dW(t)$ is the incremental Wiener process defined in section 4.2, z is the vertical coordinate, and t is time. The companion to the above equation is

$$dz = wdt. \qquad (7.2b)$$

The above equations define a continuous Markov process in velocity, space, and time.

Our problem is to define the deterministic term $a(z,w,t)$ and the stochastic coefficient $b(z,w,t)$ in Eq. (7.2a). We continue with the Fokker–Planck equation,

$$\frac{\partial P_E(z,w,t)}{\partial t} + \frac{\partial}{\partial z}(wP_E) =$$
$$-\frac{\partial}{\partial w}(aP_E) + \frac{1}{2}\frac{\partial^2}{\partial w^2}(b^2 P_E), \qquad (7.3)$$

where $P_E(z,w,t)$ is the joint unconditional probability density of the Eulerian velocity w and coordinate z at time t. The coefficients $a(z,w,t)$ and $b(z,w,t)$ have the same meanings in Eqs. (7.2a) and (7.3). Equation (7.3) comes from integrating the Fokker–Planck equation, Eq. (5.26), or the forward Kolmogorov equation, Eq. (5.29), over the probability distribution functions for the initial unbiased values of w and z [see Eq. (5.5); also Borgas and Sawford (1994b) p. 72].

The solution for the coefficient in the stochastic term, $b(z,w,t)$, is relatively simple. We make use of a relation for the Lagrangian structure function in section 3.3, Eq. (3.7), and a result in section 3.6, Eq. (3.23), to get the solution in Eq. (3.24),

$$b(z,t) = (C_0\varepsilon)^{1/2}, \qquad (7.4)$$

where C_0 is the universal constant of the Lagrangian structure function and $\varepsilon(z,t)$ is the mean rate of turbulence kinetic energy dissipation.

The following equations follow from Eq. (7.3) and define Thomson's (1987) well-mixed condition:

$$aP_E = \frac{\partial}{\partial w}\left(\frac{b^2}{2}P_E\right) + \phi(z,w,t), \qquad (7.5a)$$

$$\frac{\partial \phi}{\partial w} = -\frac{\partial P_E}{\partial t} - \frac{\partial}{\partial z}(wP_E), \qquad (7.5b)$$

and

$$\phi \to 0 \quad \text{as} \quad |w| \to \infty. \qquad (7.5c)$$

We use these relations in our solution for the coefficient $a(z,w,t)$.

We assume the probability distribution function $P_E(z,w,t)$ is Gaussian,

$$P_E(z,w,t) = (2\pi)^{-1/2}\sigma_w^{-1}\exp\left[-\frac{1}{2}\left(\frac{w}{\sigma_w}\right)^2\right],$$
$$(7.6)$$

where w is the Lagrangian velocity and $\sigma_w^2(z,t)$ is the variance of the Eulerian vertical velocity. [We noted in section 5.1 that the distinction between Eulerian and Lagrangian statistics is merely one of sampling or conditioning (Sawford 1993).] We assume that there is zero mean flow in the vertical direction. Our next step is to obtain the partial derivatives of Eq. (7.6) with respect to velocity, time, and space. We first rewrite this equation for convenience as

$$\ln(P_E) = -\frac{1}{2}\ln(2\pi) - \ln(\sigma_w) - \frac{1}{2}\left(\frac{w}{\sigma_w}\right)^2. \quad (7.7)$$

Then we get the following:

$$\frac{\partial \ln P_E}{\partial w} = -\frac{w}{\sigma_w^2}, \qquad (7.8a)$$

$$\frac{\partial \ln P_E}{\partial t} = -\frac{\partial \ln \sigma_w}{\partial t} - \frac{w^2}{2}\frac{\partial \sigma_w^{-2}}{\partial t}, \qquad (7.8b)$$

and

$$\frac{\partial \ln P_E}{\partial z} = -\frac{\partial \ln \sigma_w}{\partial z} - \frac{w^2}{2}\frac{\partial \sigma_w^{-2}}{\partial z}. \qquad (7.8c)$$

The derivatives $\partial w/\partial t$ and $\partial w/\partial z$ do not appear in the above because z, w, t are independent variables in the Eulerian Fokker–Planck equation.

It is a simple matter to get the "fading memory" component of the coefficient $a(z,w,t)$. From Eqs. (7.4), (7.5a), and (7.8a):

$$a = -\left(\frac{C_0\varepsilon}{2\sigma_w^2}\right)w + \frac{\phi}{P_E}. \qquad (7.9a)$$

The "fading memory" is represented by the first term on the right. If we combine Eq. (3.25) with the above, we get a more obvious "fading memory" term:

$$a = -\frac{w}{\tau_L} + \frac{\phi}{P_E}. \qquad (7.9b)$$

Our next step is to obtain a solution for the "drift correction" represented by the quantity ϕ/P_E. We rewrite Eq. (7.5b) as follows, remembering that z and w are independent variables:

$$\frac{1}{P_E}\frac{\partial \phi}{\partial w} = -\frac{\partial \ln P_E}{\partial t} - w\frac{\partial \ln P_E}{\partial z}.$$ (7.10)

We combine Eqs. (7.8b,c) with Eq. (7.10) to get a cubic equation in w:

$$\frac{1}{P_E}\frac{\partial \phi}{\partial w} = \frac{\partial \ln \sigma_w^2}{\partial t} +$$
$$w\frac{\partial \ln \sigma_w}{\partial z} + \frac{w^2}{2}\frac{\partial \sigma_w^{-2}}{\partial t} + \frac{w^3}{2}\frac{\partial \sigma_w^{-2}}{\partial z}.$$ (7.11)

We assume that the quantity ϕ/P_E is a quadratic function in w; that is, a function of the zeroth, first, and second moments of w:

$$\phi/P_E = \alpha + \beta w + \gamma w^2.$$ (7.12)

This is consistent with the Gaussian property of P_E (see Eq. 7.6). We differentiate Eq. (7.12) with respect to w and combine the result with Eq. (7.8a) to get another cubic function in w:

$$\frac{1}{P_E}\frac{\partial \phi}{\partial w} = \beta + \left(2\gamma - \frac{\alpha}{\sigma_w^2}\right)w - \frac{\beta}{\sigma_w^2}w^2 - \frac{\gamma}{\sigma_w^2}w^3.$$ (7.13)

We compare Eqs. (7.11) and (7.13) term by term to solve for the coefficients in Eq. (7.12). We get the identical solution for β from the first and third terms on the right side of these equations:

$$\beta = \frac{\partial \ln \sigma_w}{\partial t}.$$ (7.14)

We get an analogous solution for γ from the fourth term on the right of Eqs. (7.11) and (7.13):

$$\gamma = \frac{\partial \ln \sigma_w}{\partial z}.$$ (7.15)

The second term on the right of Eqs. (7.11) and (7.13) plus Eq. (7.15) gives the result,

$$\alpha = \sigma_w^2\frac{\partial \ln \sigma_w}{\partial z}.$$ (7.16)

Finally, we combine Eqs. (7.12) and (7.14)–(7.16) to get the solution,

$$\frac{\phi}{P_E} = \sigma_w^2\frac{\partial \ln \sigma_w}{\partial z} + w\frac{\partial \ln \sigma_w}{\partial t} + w^2\frac{\partial \ln \sigma_w}{\partial z}.$$ (7.17a)

We convert from the partial derivative of $\ln \sigma_w$ to that of the variance σ_w^2 to obtain the equivalent (assuming zero mean flow) of Eq. (30) in Thomson (1987):

$$\frac{\phi}{P_E} = \frac{1}{2}\frac{\partial \sigma_w^2}{\partial z} + \frac{1}{2\sigma_w^2}\left(\frac{\partial \sigma_w^2}{\partial t}\right)w + \frac{1}{2\sigma_w^2}\left(\frac{\partial \sigma_w^2}{\partial z}\right)w^2.$$ (7.17b)

If we assume stationary conditions in Eqs. (7.2a), (7.4), (7.9b), and (7.17b), we get Eq. (3.2), written in terms of the incremental Wiener process instead of white noise:

$$dw = -\frac{w}{\tau_L}dt +$$
$$\frac{1}{2}\left[1 + \left(\frac{w}{\sigma_w}\right)^2\right]\frac{\partial \sigma_w^2}{\partial z}dt + \left(\frac{2\sigma_w^2}{\tau_L}\right)^{1/2}dW(t).$$ (7.18)

This is the model for one-dimensional (vertical) diffusion in stationary inhomogeneous turbulence that was proposed by Wilson et al. (1983) and rigorously derived by Thomson (1984, 1987). Thomson used a different procedure each time. In section 5 of the 1984 paper, he assumed a Gaussian output of turbulent velocity fluctuations, given the first two terms on the right side of Eq. (7.18), and used moment-generating functions to determine that the forcing function (the third term on the right) must be Gaussian. In the 1987 paper, he assumed that P_E is Gaussian [see Eq. (7.6)] and used the Fokker–Planck equation, Eq. (7.3), to solve for the first two terms on the right in Eq. (7.18). The second procedure, which we used in the preceding, is less cumbersome. We use the Fokker–Planck equation in chapter 8 to derive the deterministic coefficients for the three-dimensional Langevin equation and random displacement models.

In section 6.3, we began with Eq. (7.18) and derived Eq. (6.23), the random displacement equation:

$$dz = \frac{d}{dz}(\sigma_w^2\tau)dt + (2\sigma_w^2\tau)^{1/2}dW(t).$$ (7.19)

If Eq. (7.18) meets the well-mixed criterion, the above does also (see Thomson's fifth criterion on p. 29). However, we learned in section 6.3 and stated again in the preceding section that not all Langevin models that reduce to Eq. (7.19) satisfy the well-mixed criterion.

In the next chapter, we learn that the equivalent of the above for driving the random displacement equation from the Fokker–Planck equation is to assume a constant uniform probability density function $P_{RD}(\mathbf{x}, t)$ for the random displacements. In other words, the random displacements are uniformly distributed in the phase space $\mathbf{x}$ and are independent of time. Finally, in chapter 10 on diffusion in the convective boundary layer, we use a conditional probability density function $P_{CBL}(w, z)$ that is conditioned by a uniform tracer distribution that is constant over the height of the boundary layer.

In closing, we must make it clear that imposing the well-mixed criterion *does not* restrict the solutions to well-mixed cases! Rather, this criterion ensures that the model is so formulated that the well-mixed condition is one of the *possible* solutions. One solution is that if a well-mixed tracer is passive (e.g., nonbuoyant and nonreactive) and there are no sources or sinks, the tracer will stay well mixed. Another (and more practical) solution is for an initially "clean" fluid into which a tracer is injected from a source. If the tracer is passive, the injection velocity is equal to the local fluid (mean plus fluctuation) velocity. In this case, the solution is for the evolution of the puff or plume of tracer in space and time.

Chapter 8

Turbulent Diffusion in Three Dimensions

In this chapter, we review, and in some cases expand, the material in Thomson (1987) on the probability density function (PDF) for turbulent flow, the Langevin and Fokker–Planck equations for stochastic models of turbulent diffusion, the well-mixed criterion, and solutions for the "drift" and "diffusion" coefficients in the stochastic differential equations. He began with general forms of the Lagrangian and Eulerian stochastic differential equations [Eqs. (4) and (5), respectively, in his text] and ended with the "simplest" of the possible solutions for the nonstationary, three-dimensional Langevin equation [see his Eqs. (24), (32), and the line following (32)]. There are few, if any, hints as to how his Eq. (32) was derived. The main purpose of this chapter is to follow the mathematical processes that lead from Eq. (4) toward Eq. (32), based on suggestions from Brian Sawford (1992, unpublished lecture; 1993, private communications). Our procedure is similar to that in section 7.2 for the one-dimensional case. In addition, we derive the three-dimensional random displacement model.

8.1 Multivariate Gaussian distribution

The *unconditional* probability density function for Gaussian turbulence in three dimensions given by Thomson (1987) section 2.1, p. 531 may be written, with some changes in notation, as

$$P_E = (2\pi)^{-3/2}(\det \tau_{ij})^{-1/2}$$
$$\times \exp\left[-\frac{1}{2}(u_i - U_i)\tau_{ij}^{-1}(u_j - U_j)\right]$$

(8.1)

in which

u = the Lagrangian velocity,

$U = \langle u_E \rangle$, the mean Eulerian velocity

u_E = the Eulerian velocity,

$\tau_{ij} = \langle (u_{Ei} - U_i)(u_{Ej} - U_j)\rangle$,

and

$i, j = 1, 2, 3.$

Like Thomson, we use tensor notation with the Einstein summation convention. Unlike Thomson,

we assume constant-density flow in the definitions of U and τ_{ij}. The meaning of (det τ_{ij}) is clear: it is the determinant of the tensor τ_{ij} as a 3×3 matrix. It is not explicitly stated in Thomson that τ_{ij}^{-1} is the *inverse* of τ_{ij}, not a simple reciprocal. The exponential term is one-half the product of the instantaneous velocity fluctuations and the *inverse* of the velocity covariance tensor. We learn in the following why the exponential term is written in the order $(u_i - U_i)\tau_{ij}^{-1}(u_j - U_j)$.

It is appropriate at this point to define the notation to be used in the following:

τ_{ij} = a tensor or an element of a tensor or matrix,

$[\tau_{ij}]$ = a matrix, and

$|\tau_{ij}|$ = a determinant.

The explanations or discussions in the literature are not always explicit about the inverse of the covariance tensor (or covariance matrix) in the equivalent of Eq. (8.1) for multivariate PDFs. Doob (1953) section II.3, pp. 72–73, explicitly defined a matrix as the inverse of the covariance matrix. Van Kampen (1992) used both the covariance matrix (section VIII.6, pp. 211–212) and its inverse (section I.6, p. 24) without using the words "the inverse." On the other hand, Risken (1989) section 2.2.3, pp. 23–24, clearly defined the covariance matrix and its inverse. We will adapt the cited material in Doob in discussing the following equivalent of Eq. (8.1):

$$P_E = (2\pi)^{-3/2}\lambda^{1/2}$$
$$\times \exp\left[-\frac{1}{2}\sum_{i,j=1}^{3}(u_i - U_i)[\lambda_{ij}](u_j - U_j)\right],$$

(8.2)

where

$[\lambda_{ij}] = [\tau_{ij}]^{-1}$ and

$\lambda = |\lambda_{ij}| = \det[\lambda_{ij}].$

We have shifted from tensor to matrix notation, hence the explicit summation in Eq. (8.2) as in Doob.

Doob required that the matrix $[\tau_{ij}]$ be "nonnegative definite" and "nonsingular," and stated

35

that it is also "symmetric" (i.e., $\tau_{ij} = \tau_{ji}$ where τ_{ij} is an element of the matrix). Only nonsingular matrices have inverses, and a (square) matrix is nonsingular if its determinant is nonzero [Searle (1982) section 5.5, pp. 129–130]. Therefore, $[\tau_{ij}]$ has the inverse $[\lambda_{ij}]$ if $|\tau_{ij}| \neq 0$. The inverse $[\lambda_{ij}]$ is also nonsingular, and its determinant $|\lambda_{ij}|$ is equal to the inverse of $|\tau_{ij}|$ as indicated by Eqs. (8.1) and (8.2) (Searle, section 5.6, pp. 130–131). The nonnegative property is best explained in the following context (Searle, sections 3.5–3.6, pp. 73–83). Let

$$\sum_{i,j=1}^{3} (u_i - U_i)[\lambda_{ij}](u_j - U_j) =$$

$$\sum_{i,j=1}^{3} \Delta u_i[\lambda_{ij}]\Delta u_j =$$

$$[\Delta u_1 \;\; \Delta u_2 \;\; \Delta u_3] \begin{bmatrix} \lambda_{11} & \lambda_{12} & \lambda_{13} \\ \lambda_{21} & \lambda_{22} & \lambda_{23} \\ \lambda_{31} & \lambda_{32} & \lambda_{33} \end{bmatrix} \begin{bmatrix} \Delta u_1 \\ \Delta u_2 \\ \Delta u_3 \end{bmatrix}.$$

$$(8.3)$$

The right side of Eq. (8.3) is a *quadratic form* consisting of the product of a row vector, a symmetric matrix, and a column vector. The column vector Δu_j is the transpose of the row vector Δu_i. The matrix $[\lambda_{ij}]$ is *positive definite* if the quadratic form is zero only if all $\Delta u = 0$, and it is *positive semidefinite* if the quadratic form is also zero for some $\Delta u \neq 0$. These two classes together are *nonnegative definite*. The quadratic form, with $\lambda_{ij} = \lambda_{ji}$, can be expressed as

$$\sum_{i,j} \Delta u_i[\lambda_{ij}]\Delta u_j =$$

$$\sum_i \Delta u_i^2 \lambda_{ij} + 2\sum_{j>i} \Delta u_i \Delta u_j \lambda_{ij}.$$

$$(8.4)$$

With $i, j = 1, 2, 3$, we get the nine components of a second order tensor or a 3×3 matrix. Equation (8.3) illustrates the rationale for writing the exponents as in Eqs. (8.1) and (8.2).

The above demonstrates the central role of the inverse of the kinematic Reynolds stress tensor τ_{ij} (or matrix $[\tau_{ij}]$) in the Gaussian PDF for three-dimensional turbulence. The procedure for finding the inverse of a tensor (or matrix) is given in Searle (1982) section 4.3, p. 93 and section 5.5, pp. 128–129; and Barnett (1990) section 4.1, p. 55 and section 4.3, p. 68.

8.2 Applicable Langevin and Fokker–Planck equations

It is assumed in Thomson (1987) section 2.3, pp. 533–534, that the evolution of the "marked particle" displacement and velocity $(\mathbf{x}, \mathbf{u})$ is a Markov process, and that the vectors $\mathbf{x}$ and $\mathbf{u}$ are continuous functions of time t with $d\mathbf{x}/dt = \mathbf{u}$ [see also Doob (1953) section VI.3, pp. 273–275]. Thomson's Eqs. (4a,b) are

$$du_i = a_i(\mathbf{x}, \mathbf{u}, t)dt + b_{ij}(\mathbf{x}, \mathbf{u}, t)dW_j(t) \quad (8.5a)$$

and

$$d\mathbf{x} = \mathbf{u}dt. \quad (8.5b)$$

Here

$\mathbf{x}$ = the displacement vector,

$\mathbf{u}$ = the Lagrangian velocity vector,

and

$dW_j(t)$ = an incremental Wiener process.

This process is Gaussian with

$$\langle dW_j(t) \rangle = 0 \quad (8.6a)$$

and

$$\langle dW_i(t)dW_j(s) \rangle = \delta_{ij}\delta(t - s)dtds, \quad (8.6b)$$

where δ_{ij} is the Kronecker delta and $\delta(t - s)$ is the Dirac delta function with dimension t^{-1}. The dimension of both sides of Eq. (8.6b) is t. The three-dimensional Eqs. (8.6a,b) are analogous to our one-dimensional Eqs. (4.7a,b).

We will use Thomson's Eq. (5), the forward Kolmogorov or Fokker–Planck equation,

$$\frac{\partial P_E}{\partial t} + \frac{\partial}{\partial x_i}(u_i P_E) =$$

$$-\frac{\partial}{\partial u_i}(a_i P_E) + \frac{\partial^2}{\partial u_i \partial u_j}\left(\frac{1}{2}b_{ij}b_{jk}P_E\right),$$

$$(8.7)$$

where the symbols have the same meaning as in Eqs. (8.1), (8.2), and (8.5a,b). The well-mixed condition (Thomson's section 3.1, pp. 534–535) leads to his Eqs. (9a,b) and (10), which satisfy Eqs. (8.5a,b) and (8.7) above:

$$a_i P_E = \frac{\partial}{\partial x_i}\left(\frac{1}{2}b_{ij}b_{jk}P_E\right) + \phi_i(\mathbf{x}, \mathbf{u}, t), \quad (8.8a)$$

$$\frac{\partial \phi_i}{\partial u_i} = -\frac{\partial P_E}{\partial t} - \frac{\partial}{\partial x_i}(u_i P_E), \qquad (8.8b)$$

and

$$\phi_i \rightarrow 0 \quad \text{as} \quad \mathbf{u} \rightarrow \infty. \qquad (8.9)$$

8.3 Solution for the stochastic term in the Langevin equation

Thomson used the Lagrangian structure function (see his section 4.1, p. 543, as well as our section 3.3) in combination with Eq. (8.5a) to get the result in his Eq. (24),

$$\langle b_{ij} b_{jk} \rangle = \delta_{ij} C_0 \varepsilon, \qquad (8.10)$$

for intervals greater than the Kolmogorov timescale but within the Kolmogorov inertial range (see section 3.2). Here C_0 is the universal constant associated with this structure function, and ε is the mean rate of turbulence kinetic energy dissipation with $\varepsilon(\mathbf{x}, t)$. This is the easy part of the solution for the coefficients in Eq. (8.5a). The more difficult and complex part of the solution is presented in the next section.

8.4 Solutions for the deterministic term in the Langevin equation

We begin by deriving the partial derivatives of the multivariate Gaussian distribution of Eq. (8.1). For convenience, we rewrite this equation (in tensor, not matrix notation) as

$$\ln P_E = -\frac{3}{2}\ln(2\pi)$$
$$+ \frac{1}{2}\ln \lambda - \frac{1}{2}(u_i - U_i)\lambda_{ij}(u_j - U_j),$$

$$(8.11)$$

where

     $\mathbf{u}$ = the Lagrangian velocity vector,

     $\mathbf{u}_E$ = the Eulerian velocity vector,

     $\mathbf{U} = \mathbf{U}(\mathbf{x}, t) = \langle \mathbf{u}_E \rangle$,

     $\lambda = |\lambda_{ij}|$,

     $\lambda_{ij} = \tau_{ij}^{-1}$,

     $\tau_{ij} = \tau_{ij}(\mathbf{x}, t) = \langle (u_{Ei} - U_{Ei})(u_{Ej} - U_{Ej}) \rangle$,

and

$\mathbf{x}$ and $\mathbf{u}$ are independent variables.

We differentiate Eq. (8.11) with respect to $\mathbf{x}$, $\mathbf{u}$, and t. First,

$$\frac{\partial}{\partial u_i}(\ln P_E) = -\lambda_{ik}(u_k - U_k) \qquad (8.12)$$

as shown in Searle (1982) section 12.8c, pp. 329–330, for the differential of a quadratic form. Second,

$$\frac{\partial}{\partial t}(\ln P_E) = \frac{1}{2}\frac{\partial}{\partial t}(\ln \lambda) + (u_j - U_j)\lambda_{ij}\frac{\partial U_k}{\partial t}$$
$$- \frac{1}{2}(u_j - U_j)(u_k - U_k)\frac{\partial \lambda_{jk}}{\partial t}.$$

$$(8.13)$$

Third,

$$\frac{\partial}{\partial x_i}(\ln P_E) = \frac{1}{2}\frac{\partial}{\partial x_i}(\ln \lambda) + (u_j - U_j)\lambda_{jk}\frac{\partial U_k}{\partial x_i}$$
$$- \frac{1}{2}(u_j - U_j)(u_k - U_k)\frac{\partial \lambda_{jk}}{\partial x_i}.$$

$$(8.14)$$

It is relatively simple to obtain the "fading memory" term in Thomson's solution. We get the following from Eqs. (8.8a), (8.10), and (8.12):

$$a_i = -\left(\frac{C_0 \varepsilon}{2}\right)\lambda_{ik}(u_k - U_k) + \frac{\phi_i}{P_E}. \qquad (8.15)$$

The "fading memory" term is the first term on the right. The remainder of our quest to obtain the solution for the "drift" term ϕ_i/P_E is much more difficult. It is complicated by the fact that there is no unique solution in more than one dimension (Thomson 1987, section 5.1, p. 547).

We rewrite Eq. (8.8b) as follows because x_i and u_i are independent variables:

$$\frac{1}{P_E}\frac{\partial \phi_i}{\partial u_i} = -\frac{\partial}{\partial t}(\ln P_E) - u_i\frac{\partial}{\partial x_i}(\ln P_E). \qquad (8.16)$$

We expand u_i in the second term on the right to $(u_i - U_i + U_i)$ and combine the result with Eqs. (8.13) and (8.14) to get the following:

$$\frac{1}{P_E}\frac{\partial \phi_i}{\partial u_i} = -\frac{1}{2}\left(\frac{\partial(\ln \lambda)}{\partial t} + U_i\frac{\partial(\ln \lambda)}{\partial x_i}\right) - \frac{(u_i - U_i)}{2}\frac{\partial(\ln \lambda)}{\partial x_i}$$

$$- (u_j - U_j)\lambda_{jk}\left(\frac{\partial U_k}{\partial t} + U_i\frac{\partial U_k}{\partial x_i}\right) - (u_i - U_i)(u_j - U_j)\lambda_{jk}\frac{\partial U_k}{\partial x_i}$$

$$+ \frac{(u_j - U_j)(u_k - U_k)}{2}\left(\frac{\partial \lambda_{jk}}{\partial t} + U_i\frac{\partial \lambda_{jk}}{\partial x_i}\right)$$

$$+ \frac{(u_i - U_i)(u_j - U_j)(u_k - U_k)}{2}\left(\frac{\partial \lambda_{jk}}{\partial x_i}\right).$$

$$(8.17)$$

We need a companion to Eq. (8.17). We get it by assuming that the drift term ϕ_i/P_E is a quadratic function of $(u - U)$; that is, that it is a function of the zeroth, first, and second moments of the velocity fluctuation $(u - U)$:

$$\phi_i/P_E = \alpha_i + \beta_{ij}(u_j - U_j) + \gamma_{ijk}(u_j - U_j)(u_k - U_k). \qquad (8.18)$$

As with Eq. (7.12), the preceding assumption is consistent with the Gaussian property of P_E. We substitute Eq. (8.18) into Eq. (8.12), with the subscript k in Eq. (8.12) replaced by the subscript m in order to avoid duplications in the following:

$$\frac{1}{P_E}\frac{\partial \phi_i}{\partial u_i} = \beta_{ii} + \gamma_{iik}(u_k - U_k) + \gamma_{iji}(u_j - U_j) - \alpha_i\lambda_{im}(u_m - U_m)$$

$$- \beta_{ij}\lambda_{im}(u_j - U_j)(u_m - U_m) - \gamma_{ijk}\lambda_{im}(u_j - U_j)(u_k - U_k)(u_m - U_m).$$

$$(8.19)$$

This is the companion to Eq. (8.17). We equate the terms of same order in Eqs. (8.17) and (8.19) to get the following four equations:

$$\beta_{ii} = -\frac{1}{2}\left(\frac{\partial(\ln \lambda)}{\partial t} + U_i\frac{\partial(\ln \lambda)}{\partial x_i}\right), \qquad (8.20a)$$

$$\alpha_i\lambda_{im}(u_m - U_m) - \gamma_{iik}(u_k - U_k) - \gamma_{iji}(u_j - U_j) =$$

$$\frac{(u_i - U_i)}{2}\frac{\partial(\ln \lambda)}{\partial x_i} + (u_j - U_j)\lambda_{jk}\left(\frac{\partial U_k}{\partial t} + U_i\frac{\partial U_k}{\partial x_i}\right), \qquad (8.20b)$$

$$\beta_{ij}\lambda_{im}(u_j - U_j)(u_m - U_m) =$$

$$(u_i - U_i)(u_j - U_j)\lambda_{jk}\frac{\partial U_k}{\partial x_i} - \frac{(u_j - U_j)(u_k - U_k)}{2}\left(\frac{\partial \lambda_{jk}}{\partial t} + U_i\frac{\partial \lambda_{jk}}{\partial x_i}\right), \qquad (8.20c)$$

and

$$\gamma_{ijk}\lambda_{im}(u_j - U_j)(u_k - U_k)(u_m - U_m) = -\frac{(u_j - U_j)(u_k - U_k)(u_m - U_m)}{2}\left(\frac{\partial \lambda_{jk}}{\partial x_m}\right). \qquad (8.20d)$$

Our next task is to solve for α_i, β_{ij}, and γ_{ijk}. There are multiple solutions for each, which is why there is no unique solution for the drift term in more than one dimension. It may be that there are additional physical or mathematical criteria (unknown at present) that would give a unique solution. It is simplest to begin with Eq. (8.20d), the cubic equation in $(u-U)$. We let $i \to m$ on the right side of this equation to get

$$\gamma_{ijk}\lambda_{im}(u_j - U_j)(u_k - U_k)(u_m - U_m) = \\ -\frac{(u_i - U_i)(u_j - U_j)(u_k - U_k)}{2}\left(\frac{\partial\lambda_{jk}}{\partial x_i}\right).$$

(8.21)

The coefficients of the cubic terms on both sides of Eq. (8.21) are equal for all combinations of $j, k, m = 1, 2, 3$ so

$$\gamma_{ijk} = -\frac{\lambda_{im}^{-1}}{2}\frac{\partial\lambda_{jk}}{\partial x_m}.$$

(8.22a)

This is the first of three solutions for γ_{ijk}. The second comes from the cyclic permutation of the indices $j \to k \to m \to j$ on the right side of Eq. (8.21):

$$\gamma_{ijk} = -\frac{\lambda_{im}^{-1}}{2}\frac{\partial\lambda_{km}}{\partial x_j}.$$

(8.22b)

The third, and last, comes from a further cyclic permutation of the same indices on the same side of Eq. (8.21):

$$\gamma_{ijk} = -\frac{\lambda_{im}^{-1}}{2}\frac{\partial\lambda_{mj}}{\partial x_k}.$$

(8.22c)

We continue with the quadratic equation in $(u-U)$, Eq. (8.20c). We change $i \to m$ in the first term on the right and $k \to m$ in the second in order to get both sides in terms of $(u_j - U_j)(u_m - U_m)$. In addition, to avoid future difficulty, we let $i \to k$ in the second term on the right. The result is

$$\beta_{ij}\lambda_{im}(u_j - U_j)(u_m - U_m) = \\ (u_j - U_j)(u_m - U_m) \\ \left[\lambda_{jk}\frac{\partial U_k}{\partial x_m} - \frac{1}{2}\left(\frac{\partial\lambda_{jm}}{\partial t} + U_k\frac{\partial\lambda_{jm}}{\partial x_k}\right)\right].$$

(8.23)

The coefficients of the quadratic terms on both sides of the above must be equal for all combinations of

$j, m = 1, 2, 3$. The first solution for β_{ij} is obtained by multiplying both sides by λ_{mn}. We make use of the Kronecker delta δ_{ij} as a second-order tensor (Jaeger, 1966, section 3.6) to make the transformation

$$\beta_{ij}\lambda_{im}\lambda_{mn}^{-1} = \beta_{ij}\,\delta_{in} = \beta_{nj}.$$

(8.24)

We substitute Eq. (8.24) into Eq. (8.23), let $n = i$ on both sides of the result, and use the relation $\lambda_{im}^{-1} = \tau_{im}$ as in Eqs. (8.2) and (8.11) to get the first result for β_{ij}:

$$\beta_{ij} = \lambda_{im}^{-1}\frac{\partial U_k}{\partial x_m}\lambda_{jk} - \frac{\tau_{im}}{2}\left(\frac{\partial\lambda_{jm}}{\partial t} + U_k\frac{\partial\lambda_{jm}}{\partial x_k}\right).$$

(8.25a)

The second solution results from the cyclic permutation $j \to m \to j$ on the right side of Eq. (8.23):

$$\beta_{ij} = \frac{\partial U_i}{\partial x_j} - \frac{\tau_{im}}{2}\left(\frac{\partial\lambda_{jm}}{\partial t} + U_j\frac{\partial\lambda_{jm}}{\partial x_k}\right).$$

(8.25b)

The second term on the right of Eq. (8.25b) is unchanged from that in Eq. (8.25a) because $\lambda_{jm} = \lambda_{mj}$. The change in the first term on the right is a consequence of a transformation similar to that in Eq. (8.24).

We now consider Eq. (8.20b), which is linear in $(u-U)$. On the left, we let $m \to j$ in the first term and $k \to j$ in the second. On the right, we let $i \to j$ in the first term and, to avoid future difficulty, $i \to m$ in the second. The result is

$$(\alpha_i\lambda_{ij} - \gamma_{iij} - \gamma_{iji})(u_j - U_j) = \\ \left[\frac{1}{2}\frac{\partial(\ln\lambda)}{\partial x_j} + \lambda_{jk}\left(\frac{\partial U_k}{\partial t} + U_m\frac{\partial U_k}{\partial x_m}\right)\right](u_j - U_j).$$

(8.26)

The coefficient of $(u_j - U_j)$ on the left must be equal to that on the right for all values of $j = 1, 2, 3$, so we equate these coefficients:

$$\alpha_i\lambda_{ij} - \gamma_{kkj} - \gamma_{kjk} = \\ \frac{1}{2}\frac{\partial(\ln\lambda)}{\partial x_j} + \lambda_{jk}\left(\frac{\partial U_k}{\partial t} + U_m\frac{\partial U_k}{\partial x_m}\right).$$

(8.27)

Note that we let $i \rightarrow k$ in the second and third terms on the left in order to avoid future difficulty. We move the terms with γ_{kkj} and γ_{kjk} to the right, multiply both sides by λ_{jm}^{-1}, apply the principles of Eq. (8.24) to simplify terms, use the relation $\lambda_{ij}^{-1} = \tau_{ij}$, and let $m \rightarrow j$ within the last term to get the following:

$$\alpha_i = (\gamma_{kkj} + \gamma_{kjk})\lambda_{ij}^{-1}$$
$$+ \frac{\tau_{ij}}{2}\frac{\partial(\ln\lambda)}{\partial x_j} + \left(\frac{\partial U_i}{\partial t} + U_j\frac{\partial U_i}{\partial x_j}\right).$$

$$(8.28)$$

We have changed the sequence of the components of the first term on the right in going from Eq. (8.28) to Eq. (8.29) in preparation for the following. We note that there is more than one solution for α_i because of the multiple solutions in Eqs. (8.22a–c) for γ_{ijk}. From Eq. (8.22a) we get the first solution for the first term in Eq. (8.28):

$$(\gamma_{kkj} + \gamma_{kjk})\lambda_{ij}^{-1} = -\lambda_{jm}^{-1}\frac{\partial\lambda_{jk}}{\partial x_m}\lambda_{ij}^{-1} = \frac{\partial\lambda_{im}^{-1}}{\partial x_m}.$$
$$(8.29a)$$

The second solution comes from *both* Eq. (8.22b) and Eq. (8.22c):

$$(\gamma_{kkj} + \gamma_{kjk})\lambda_{ij}^{-1} = \frac{1}{2}\left(\frac{\partial\lambda_{ik}}{\partial x_k} - \lambda_{ij}^{-1}\frac{\partial(\ln\lambda)}{\partial x_j}\right).$$
$$(8.29b)$$

Let us explain these results. First, for the procedure in Eq. (8.29a) and similar parts of Eq. (8.29b):

$$\frac{\partial(\lambda_{jk}\lambda_{ij}^{-1})}{\partial x_m} = \frac{\partial\delta_{ik}}{\partial x_m} = 0 = \lambda_{jk}\frac{\partial\lambda_{ij}^{-1}}{\partial x_m} + \lambda_{ij}^{-1}\frac{\partial\lambda_{jk}}{\partial x_m}.$$
$$(8.30a)$$

From the right side of Eq. (8.30a) multiplied by λ_{km}^{-1}:

$$-\lambda_{km}^{-1}\frac{\partial\lambda_{jk}}{\partial x_m}\lambda_{ij}^{-1} = \lambda_{km}^{-1}\frac{\partial\lambda_{ij}^{-1}}{\partial x_m}\lambda_{jk} =$$
$$\delta_{jm}\frac{\partial\lambda_{ij}^{-1}}{\partial x_m} = \frac{\partial\lambda_{im}^{-1}}{\partial x_m}.$$
$$(8.30b)$$

The above in Eqs. (8.30a,b) is equivalent to Searle (1982) section 12.10.c, p. 335. Material in section 12.10.d, pp. 337–338, is applicable to the other parts of the solution common to the two paths that lead from Eqs. (8.22b,c) to Eq. (8.29b). In our tensor notation (see also section 2.8.k, pp. 45–46, in Searle), the differential of a determinant may be written as

$$\frac{\partial(\ln\lambda)}{\partial x_j} = \lambda_{km}^{-1}\frac{\partial\lambda_{km}}{\partial x_j}.$$
$$(8.31)$$

With $k, m = 1, 2, 3$, there are nine terms in the expansion of the right side of Eq. (8.31). We now have all the components we need for the solution for ϕ_i/P_E: Eqs. (8.18), (8.22a–c), (8.25a,b), (8.28), and (8.29a,b).

For the sake of completeness, let us consider the relation between Eqs. (8.25a,b) for β_{ij} and Eq. (8.20a) for β_{ii}. If we let $\beta_{ij} \rightarrow \beta_{ii}$, Eqs. (8.25a,b) reduce to the following if we apply the constant-density flow condition $\partial U_i/\partial x_i = 0$:

$$\beta_{ii} = -\frac{\lambda_{im}^{-1}}{2}\left(\frac{\partial\lambda_{im}}{\partial t} + U_k\frac{\partial\lambda_{im}}{\partial x_k}\right).$$
$$(8.32)$$

This is equivalent to Eq. (8.20a) and is consistent with Eq. (8.31).

8.5 The Borgas solution and Thomson's "simplest" solution

In the preceding solutions for ϕ_i/P_E as defined by Eq. (8.18), we get three solutions for γ_{ijk}, two solutions for β_{ij}, and two solutions for α_i. All are mathematically valid. An unanswered question at the time of writing is, "Which combination of these solutions is closest to physical reality?" Let us begin with the first solution for γ_{ijk}, Eq. (8.22a), the first solution for β_{ij}, Eq. (8.25a), and the first solution for α_i, a combination of Eqs. (8.28) and (8.29a). The result is, after making some use of the relation $\tau_{ij} = \lambda_{ij}^{-1}$ and changing a few indices,

$$\frac{\phi_i}{P_E} = \frac{\partial U_i}{\partial t} + U_j\frac{\partial U_i}{\partial x_j} + \frac{\partial\tau_{ij}}{\partial x_j} + \frac{\tau_{im}}{2}\frac{\partial(\ln\lambda)}{\partial x_m}$$
$$+ \left[\lambda_{im}^{-1}\frac{\partial U_k}{\partial x_m}\lambda_{jk} - \frac{\tau_{im}}{2}\left(\frac{\partial\lambda_{jm}}{\partial t} + U_k\frac{\partial\lambda_{jm}}{\partial x_k}\right)\right]$$
$$\times (u_j - U_j)$$
$$- \left[\frac{\tau_{im}}{2}\frac{\partial\lambda_{jm}}{\partial x_j}\right](u_j - U_j)(u_k - U_k).$$

$$(8.33)$$

This result is equivalent to the solution obtained by Michael Borgas [Eq. (10) in Sawford and Guest 1988]. We continue with the second solution for γ_{ijk} in Eq. (8.22b), the second solution for β_{ij} in

Eq. (8.25b), and the second solution for α_i, a combination of Eqs (8.28) and (8.29b). The result is

$$
\frac{\phi_i}{P_E} = \frac{\partial U_i}{\partial t} + U_j \frac{\partial U_i}{\partial x_j} + \frac{1}{2} \frac{\partial \tau_{ij}}{\partial x_j}
$$
$$
+ \left[\frac{\partial U_i}{\partial x_j} - \frac{\tau_{im}}{2} \left(\frac{\partial \lambda_{jm}}{\partial t} + U_k \frac{\partial \lambda_{jm}}{\partial x_k} \right) \right] (u_j - U_j)
$$
$$
- \left[\frac{\tau_{im}}{2} \frac{\partial \lambda_{km}}{\partial x_j} \right] (u_j - U_j)(u_k - U_k).
$$
$$(8.34)$$

The above is equivalent to Eq. (32) on p. 547 in Thomson (1987) and Eq. (9) in Sawford and Guest (1988). The form of the above two equations is like that in Sawford and Guest.

We subtract Eq. (8.34) from Eq. (8.33) to illustrate the difference between the Borgas and Thomson solutions:

$$
\left(\frac{\phi_i}{P_E} \right)_B - \left(\frac{\phi_i}{P_E} \right)_T = \frac{1}{2} \frac{\partial \tau_{ij}}{\partial x_j}
$$
$$
+ \frac{\tau_{im}}{2} \frac{\partial (\ln \lambda)}{\partial x_m} + \left[\lambda_{im}^{-1} \frac{\partial U_k}{\partial x_m} \lambda_{jk} - \frac{\partial U_i}{\partial x_j} \right] (u_j - U_j).
$$
$$(8.35)$$

This shows that the difference is in the coefficients α_i and β_{ij} in Eq. (8.18) for ϕ_i/P_E.

We rephrase the question we asked at the beginning of this section: "Which solution is closer to physical reality?" We have no answer at present, but one of the basic philosophical rules in science is to favor simplicity; for example, the Copernican model of the solar system rather than the Ptolemaic. Therefore, we will use Thomson's "simplest" solution, Eq. (8.34), in subsequent chapters.

8.6 Solutions for the random displacement and diffusion equations

Let us use an analogous procedure to solve for the coefficients a_i and b_{ij} in the three-dimensional random displacement equation (B. Sawford 1992, unpublished lecture notes):

$$
dx_i = a_i(\mathbf{x}, t)dt + b_{ij}(\mathbf{x}, t)dW_j(t), \qquad (8.36)
$$

where the symbols $\mathbf{x}$ and $dW(t)$ mean the same as in Eqs. (8.5)–(8.6). The corresponding Fokker–Planck equation for the probability density function for the displacement $P_{RD}(\mathbf{x}, t)$ is

$$
\frac{\partial P_{RD}}{\partial t} = -\frac{\partial (a_i P_{RD})}{\partial x_i} + \frac{1}{2} \frac{\partial^2 (b_{ik} b_{kj} P_{RD})}{\partial x_i \partial x_j}. \quad (8.37)
$$

For the random displacement model, the equivalent of the well-mixed condition for the Langevin equation model is

$$
P_{RD}(\mathbf{x}, t) = \text{constant}. \qquad (8.38)
$$

The above means that the well-mixed condition is *one* of the possible solutions of Eq. (8.37). It *does not* preclude P_{RD} from being variable in time and space. The combination of Eqs. (8.37) and (8.38) gives the result,

$$
\frac{\partial a_i}{\partial x_i} = \frac{1}{2} \frac{\partial^2 (b_{ik} b_{kj})}{\partial x_i \partial x_j}. \qquad (8.39)
$$

The solution is

$$
a_i = \frac{1}{2} \frac{\partial (b_{ij} b_{kj})}{\partial x_i} + \phi_i \qquad (8.40a)
$$

with

$$
\frac{\partial \phi_i(\mathbf{x}, t)}{\partial x_i} = 0. \qquad (8.40b)
$$

If we assume constant-density flow, we can identify the mean Eulerian velocity $U_i(\mathbf{x}, t)$ with $\phi_i(\mathbf{x}, t)$ in Eqs. (8.40a,b):

$$
\phi_i(\mathbf{x}, t) = U_i(\mathbf{x}, t). \qquad (8.41a)
$$

In addition, we let

$$
\langle b_{ik} b_{kj} \rangle = 2K_{ij}(\mathbf{x}, t), \qquad (8.41b)
$$

where K_{ij} is the eddy diffusivity coefficient. We get the following from Eqs. (8.40a,b) and (8.41a,b):

$$
a_i(\mathbf{x}, t) = \frac{\partial K_{ij}(\mathbf{x}, t)}{\partial x_i} + U_i(\mathbf{x}, t), \qquad (8.42a)
$$

and

$$
\frac{\partial U_i(\mathbf{x}, t)}{\partial x_i} = 0. \qquad (8.42b)
$$

The solution for the random displacement equation, Eq. (8.36), is

$$dx_i = \left[U_i(\mathbf{x}, t) + \frac{\partial K_{ij}(\mathbf{x}, t)}{\partial x_j} \right] dt$$
$$+ (2K_{ij}(\mathbf{x}, t))^{1/2} dW_j(t).$$

$$(8.43)$$

Unlike the case of the three-dimensional Langevin model of turbulent diffusion, physics [Eqs. (8.40b) and (8.41a)] leads to a unique solution for the three-dimensional random displacement model. In closing, let us solve for the equivalent diffusion equation. We begin by substituting Eqs (8.41b) and (8.42a,b) into Eq. (8.37), with the subscript in P_{RD}

dropped for convenience:

$$\frac{\partial P}{\partial t} = -U_i \frac{\partial P}{\partial x_i}$$
$$- \frac{\partial}{\partial x_i} \left(P \frac{\partial K_{ij}}{\partial x_j} \right) + \frac{\partial}{\partial x_i} \left[\frac{\partial}{\partial x_j} (K_{ij} P) \right].$$

$$(8.44a)$$

The above equation reduces to the familiar form of the Eulerian advection–diffusion equation (we have substituted the tracer concentration C for the probability density function P):

$$\frac{\partial C}{\partial t} + U_i \frac{\partial C}{\partial x_i} = \frac{\partial}{\partial x_i} \left(K_{ij} \frac{\partial C}{\partial x_j} \right). \qquad (8.44b)$$

Chapter 9

Applications of Thomson's "Simplest" Solution

The theme of this chapter is the simplification of Thomson's (1987) solution for the nonstationary, three-dimensional Langevin equation model for turbulent diffusion. For our convenience, we repeat Eqs. (8.8a,b), (8.10), and (8.15):

$$a_i P_E = \frac{\partial}{\partial x_i}\left(\frac{1}{2}b_{ij}b_{jk}P_E\right) + \phi_i(\mathbf{x},\mathbf{u},t), \quad (9.1\text{a})$$

$$\frac{\partial \phi_i}{\partial u_i} = -\frac{\partial P_E}{\partial t} - \frac{\partial}{\partial x_i}(u_i P_E), \quad (9.1\text{b})$$

$$\langle b_{ij}b_{jk}\rangle = \delta_{ij}C_0\varepsilon, \quad (9.2)$$

$$a_i = -\left(\frac{C_0\varepsilon}{2}\right)\lambda_{ik}(u_k - U_k) + \frac{\phi_i}{P_E}, \quad (9.3)$$

and Thomson's "simplest" solution,

$$\frac{\phi_i}{P_E} = \frac{\partial U_i}{\partial t} + U_j\frac{\partial U_i}{\partial x_j} + \frac{1}{2}\frac{\partial \tau_{ij}}{\partial x_j}$$
$$+ \left[\frac{\partial U_i}{\partial x_j} - \frac{\tau_{im}}{2}\left(\frac{\partial \lambda_{jm}}{\partial t} + U_k\frac{\partial \lambda_{jm}}{\partial x_k}\right)\right](u_j - U_j)$$
$$- \left[\frac{\tau_{im}}{2}\frac{\partial \lambda_{km}}{\partial x_j}\right](u_j - U_j)(u_k - U_k). \quad (9.4)$$

9.1 Simplification for three dimensions

The number of indices in the terms on the right of Eq. (9.4) range from one to four. This means that, in the general case, the function ϕ_i/P_E has a total of 73 components for each index i. In other words, with $i = i$ and summation over j, k, m, the right side of Eq. (9.4) has a total of 73 terms in its expanded form [we retain the velocity differences $(u_j - U_j)$ and $(u_k - U_k)$]. It is clear that simplification is essential if we are to use the above model in practical applications. We begin by changing from the partial derivatives of the inverse tensor λ_{ij} to those of the kinematic Reynolds stress tensor τ_{ij}:

$$\lambda_{ij} = \tau_{ij}^{-1} \quad (9.5\text{a})$$

so

$$\tau_{ij}\frac{\partial \lambda_{ij}}{\partial t} = -\lambda_{ij}\frac{\partial \tau_{ij}}{\partial t} \quad \text{and} \quad \tau_{ij}\frac{\partial \lambda_{ij}}{\partial x_k} = -\lambda_{ij}\frac{\partial \tau_{ij}}{\partial x_k}. \quad (9.5\text{b})$$

We make this change because it will be easier to obtain partial derivatives of τ_{ij} than of its inverse λ_{ij}. This will also put Eq. (9.4) in the form given in Thomson (1987):

$$\frac{\phi_i}{P_E} = \frac{\partial U_i}{\partial t} + U_j\frac{\partial U_i}{\partial x_j} + \frac{1}{2}\frac{\partial \tau_{ij}}{\partial x_j}$$
$$+ \left[\frac{\partial U_i}{\partial x_j} + \frac{\lambda_{im}}{2}\left(\frac{\partial \tau_{jm}}{\partial t} + U_k\frac{\partial \tau_{jm}}{\partial x_k}\right)\right](u_j - U_j)$$
$$+ \left[\frac{\lambda_{im}}{2}\frac{\partial \tau_{km}}{\partial x_j}\right](u_j - U_j)(u_k - U_k). \quad (9.6)$$

We assume turbulent flow that is stationary $((\partial/\partial t) = 0)$ and homogeneous in the horizontal plane $((\partial/\partial x_1 = \partial/\partial x_2 = 0$ or $\partial/\partial x = \partial/\partial y = 0)$ with zero mean lateral and vertical flow $(U_2 = U_3 = 0$ or $V = W = 0)$. The result is turbulent flow that can be vertically inhomogeneous $(\partial/\partial x_3 \neq 0$ or $\partial/\partial z \neq 0)$ and is aligned with the longitudinal $(x_1$ or $x)$ coordinate and flow $(U_1 \neq 0$ or $U \neq 0)$. Under these conditions, it can be shown (see, e.g., Calder 1966) that the Reynolds stress tensor reduces to

$$\tau_{ij} = \begin{bmatrix} \tau_{11} & 0 & \tau_{13} \\ 0 & \tau_{22} & 0 \\ \tau_{31} & 0 & \tau_{33} \end{bmatrix} \quad (9.7\text{a})$$

with

$$\tau_{13} = \tau_{31} \quad (9.7\text{b})$$

for a total of only four components to define τ_{ij}.

The resulting stochastic differential equations for the three Lagrangian velocity components are, from Eqs. (9.1)–(9.3) and (9.6)–(9.7),

$$du_1 = \left[-\frac{C_0\varepsilon}{2}[\lambda_{11}(u_1 - U_1) + \lambda_{13}u_3] + \frac{\partial U_1}{\partial x_3}u_3 + \frac{1}{2}\frac{\partial \tau_{13}}{\partial x_3} \right] dt$$

$$\left[\frac{\partial \tau_{11}}{\partial x_3}[\lambda_{11}(u_1 - U_1) + \lambda_{13}u_3] + \frac{\partial \tau_{13}}{\partial x_3}[\lambda_{13}(u_1 - U_1) + \lambda_{33}u_3] \right] \frac{u_3}{2}dt$$

$$+ (C_0\varepsilon)^{1/2}dW_1(t),$$

(9.8a)

$$du_2 = \left[-\frac{C_0\varepsilon}{2}(\lambda_{22}u_2) + \frac{\partial \tau_{22}}{\partial x_3}(\lambda_{22}u_2)\frac{u_3}{2} \right] dt + (C_0\varepsilon)^{1/2}dW_2(t),$$

(9.8b)

and

$$du_3 = \left[-\frac{C_0\varepsilon}{2}[\lambda_{13}(u_1 - U_1) + \lambda_{33}u_3] + \frac{1}{2}\frac{\partial \tau_{33}}{\partial x_3} \right] dt$$

$$+ \left[\frac{\partial \tau_{13}}{\partial x_3}[\lambda_{11}(u_1 - U_1) + \lambda_{13}u_3] + \frac{\partial \tau_{33}}{\partial x_3}[\lambda_{13}(u_1 - U_1) + \lambda_{33}u_3] \right] \frac{u_3}{2}dt$$

$$+ (C_0\varepsilon)^{1/2}dW_3(t).$$

(9.8c)

The components of the inverse of the tensor τ_{ij}, λ_{ij}, are as follows [see Eq. (8.9)]:

$$\lambda_{11} = (\tau_{11} - \tau_{13}^2/\tau_{33})^{-1}, \qquad (9.9a)$$

$$\lambda_{22} = \tau_{22}^{-1}, \qquad (9.9b)$$

$$\lambda_{33} = (\tau_{33} - \tau_{13}^2/\tau_{11})^{-1}, \qquad (9.9c)$$

and

$$\lambda_{13} = (\tau_{13} - \tau_{11}\tau_{33}/\tau_{13})^{-1}. \qquad (9.9d)$$

Note that λ_{22} is equal to the simple reciprocal of the Reynolds stress τ_{22}. Also note that if $\tau_{13} \to 0$, $\lambda_{11} \to \tau_{11}^{-1}$ and $\lambda_{33} \to \tau_{33}^{-1}$. There is coupling between pairs of Eqs. (9.8a–c): between u_1 and u_3 in Eqs. (9.8a) and (9.8c), and u_2 is a function of u_3 in Eq. (9.8b) but u_3 is not a function of u_2.

For convenience, we define the components of the Reynolds stress tensor τ_{ij} in terms of more familiar symbols: $\tau_{11} = \sigma_u^2$, $\tau_{22} = \sigma_v^2$, $\tau_{33} = \sigma_w^2$, and $\tau_{13} = \tau_{uw}$. In the preceding, the first three components are kinematic normal Reynolds stresses and the fourth is a kinematic Reynolds shear stress.

9.2 Simplification to two dimensions and to one dimension

For turbulent diffusion in only the lateral and vertical dimensions (x_2 and x_3 or y and z), we consider only the stochastic differential equations for du_2 and du_3 (or dv and dw). Equation (9.8b) is unchanged; we repeat it here for convenience after

making use of Eq. (9.9b):

$$du_2 = \left[-\frac{C_0\varepsilon}{2}\left(\frac{u_2}{\tau_{22}}\right) + \frac{1}{2}\left(\frac{u_2u_3}{\tau_{22}}\right)\frac{\partial \tau_{22}}{\partial x_3} \right] dt$$

$$+ (C_0\varepsilon)^{1/2}dW_2(t).$$

(9.10a)

Equation (9.8c) is drastically changed because the quantities $u_1 - U_1$ and τ_{13} associated with longitudinal diffusion and the mean velocity gradient $\partial U_1/\partial x_3$ vanish. We use Eq. (9.9c) with $\tau_{13} = 0$ to get

$$du_3 = \left[-\frac{C_0\varepsilon}{2}\left(\frac{u_3}{\tau_{33}}\right) + \frac{1}{2}\left(1 + \frac{u_3^2}{\tau_{33}}\right)\frac{\partial \tau_{33}}{\partial x_3} \right] dt$$

$$+ (C_0\varepsilon)^{1/2}dW_3(t).$$

(9.10b)

Equation (9.10b) is a companion in two dimensions to Eq. (9.10a). In addition, it is equivalent to the combination of Eqs. (3.25) and (7.18) for turbulent diffusion in one dimension. It is an alternate form of the one-dimensional model originally proposed by Wilson et al. (1983) and derived rigorously (using two different methods) by Thomson (1984, 1987).

Chapter 10

Application to the Convective Boundary Layer

For most applications, the assumption of Gaussian turbulence in deriving the several stochastic Lagrangian models in chapters 6–9 is an acceptable approximation. A major exception is the convective boundary layer (CBL) because the persistent updrafts and downdrafts dominate the vertical flow. The resultant trajectory of a passive tracer cannot, therefore, be regarded as a superposition of independent increments as in Eqs. (9.8c) and (9.10b), for example. For this reason the Eulerian eddy diffusivity model, Eq. (8.44b), is not applicable *everywhere* in the CBL. For an area source at the top of a horizontally homogeneous CBL, the effective eddy diffusivity, K_{eff}, is well-behaved for downward diffusive transport. On the other hand, the eddy diffusivity model fails for an area source at the bottom of the same CBL. There is a singularity in the value of K_{eff} (from $K_{eff} \gg 0$ to $K_{eff} \ll 0$) about halfway up and a countergradient flux (meaning negative K_{eff}) in the upper part of the CBL. The situation is more complex for vertical diffusion with horizontal advection from a steady source (point or horizontal line) near the surface in the CBL. In this case, K_{eff} is ill-behaved (negative) in some portions of the plume (for example, see the appendix in Deardorff and Willis 1975). Wyngaard (1987) proposed that the skewness of the vertical velocity field is responsible for this diffusion phenomenon. Numerical simulations by Wyngaard and Weil (1991) suggest that this transport asymmetry is caused by the interaction between the skewness of the turbulence and the gradient of the scalar flux. There have been attempts to reformulate this model for use in the CBL. For example, Holtslag and Moeng (1991) proposed a modification that includes countergradient terms and separate parameterizations for top-down and bottom-up diffusion. Their model is well behaved over most of the CBL, but gives negative values of K_{eff} in the upper part of the CBL under some conditions. This failure of the eddy diffusivity model [and the equivalent Lagrangian random displacement model, Eq. (6.23)] provides the principal motivation for applying the Lagrangian Langevin equation model to the CBL.

10.1 Approximating non-Gaussian turbulence

The PDF for the vertical velocity fluctuations in the CBL is of special importance in developing a Langevin model for turbulent diffusion. The PDF of the CBL is non-Gaussian because of the organized motion consisting of updrafts and downdrafts. The PDF has a negative mode and a positive tail such that the third moment (or skewness) is positive. This is because updrafts have higher velocities but occupy less area than downdrafts.

Baerentsen and Berkowicz (1984), referred to hereafter in many cases as BB, made the valuable contribution of approximating the observed non-Gaussian PDF of the CBL with the sum of two Gaussian distributions, one for the updrafts and the other for the downdrafts. They required that certain relations for the zeroth through third moments of the PDF be satisified. This technique was used by de Baas et al. (1986), Sawford and Guest (1987), Luhar and Britter (1989), and Weil (1990). Du et al. (1994) extended this technique to include the fourth moment. Luhar et al. (1996) reviewed the preceding procedures, discussed some of their limitations, and proposed an alternate.

Baerentsen and Berkowicz introduced a skewed PDF consisting of the sum of two Gaussian distributions as follows:

$$P_{CBL}(w|z) = \frac{1}{\sqrt{2\pi}} \frac{A_1}{\sigma_1} \exp\left[-\frac{1}{2}\left(\frac{w - m_1}{\sigma_1}\right)^2\right]$$
$$+ \frac{1}{\sqrt{2\pi}} \frac{A_2}{\sigma_2} \exp\left[-\frac{1}{2}\left(\frac{w - m_2}{\sigma_2}\right)^2\right],$$

$$(10.1)$$

where A_1 and A_2 are the respective probabilities of occurrence of updrafts and downdrafts. They may also be interpreted as the respective areas occupied by updrafts and downdrafts. Here the mean updraft and downdraft velocities are

$$m_1 = \langle w_1 \rangle \quad \text{and} \quad m_2 = \langle w_2 \rangle, \qquad (10.2a)$$

and the standard deviations of the velocities in the updrafts and downdrafts are

$$\sigma_1 = (\langle w_1^2 \rangle)^{1/2} \quad \text{and} \quad \sigma_2 = (\langle w_2^2 \rangle)^{1/2}. \qquad (10.2b)$$

We choose m_1 for the updrafts to be positive and m_2 for the downdrafts to be negative. The parameters A, m, σ are functions of the elevation z. For a given z, $P_{CBL}(w|z)$ is defined by six parameters: $A_1, A_2, m_1, m_2, \sigma_1, \sigma_2$. Luhar et al. (1996) stated that, in principle, the values of these six unknowns can be obtained by using

$$\langle w^n \rangle = \int_{-\infty}^{\infty} w^n P_{CBL}(w|z) dw \qquad (10.3)$$

to obtain (with $n = 0, 1, 2, 3, 4, 5$) the following for the zeroth through the fifth moments of the PDF:

$$A_1 + A_2 = 1, \qquad (10.4a)$$

$$A_1 m_1 + A_2 m_2 = 0, \qquad (10.4b)$$

$$A_1(m_1^2 + \sigma_1^2) + A_2(m_2^2 + \sigma_2^2) = \sigma_w^2, \qquad (10.4c)$$

$$A_1(m_1^3 + 3m_1\sigma_1^2) + A_2(m_2^3 + 3m_2\sigma_2^2) = \langle w^3 \rangle, \qquad (10.4d)$$

$$A_1(m_1^4 + 6m_1^2\sigma_1^2 + 3\sigma_1^4) + A_2(m_2^4 + 6m_2^2\sigma_2^2 + 3\sigma_2^4) = \langle w^4 \rangle, \qquad (10.4e)$$

$$A_1(m_1^5 + 10m_1^3\sigma_1^2 + 15m_1\sigma_1^4) + A_2(m_2^5 + 10m_2^3\sigma_2^2 + 15m_2\sigma_2^4) = \langle w^5 \rangle. \qquad (10.4f)$$

As Luhar et al. noted, there are two problems with using the above set of six simultaneous equations to obtain values of the six parameters

$$A_1, A_2, m_1, m_2, \sigma_1, \sigma_2.$$

First, values of the zeroth through fifth moments of w are required for a solution, and little or no information is generally available on the fourth and fifth moments. Second, even if values for all the moments would be available, only numerical solutions of Eqs. (10.4a–f) would be possible. This would

be computationally undesirable in a simulation requiring thousands of time steps for each of tens of thousands of particle trajectories. In the following, we describe more efficient closures in which simple relations are used instead of one or two of the higher-moment equations.

We begin with the technique of Baerentsen and Berkowicz (1984), as modified by Weil (1990), in which only Eqs. (10.4a–d) are used. We combine Eqs. (10.4a,b) to get solutions for A_1 and A_2:

$$A_1 = m_2(m_2 - m_1)^{-1}, \qquad (10.5a)$$

$$A_2 = -m_1(m_2 - m_1)^{-1}. \qquad (10.5b)$$

We introduce the ratio R and assume the ratios for the updrafts and downdrafts are equal (Weil 1990):

$$R = \sigma_1/m_1 = -\sigma_2/m_2. \qquad (10.6)$$

[Weil assumed $R = 3/2$ but de Baas et al. (1986), Sawford and Guest (1987), and Luhar and Britter (1989) followed BB in assuming $R = 1$.] We also make use of the skewness S,

$$S = \langle w^3 \rangle/\sigma_w^3. \qquad (10.7)$$

For convenience Weil introduced

$$\alpha = (1 + R^2)(1 + 3R^2)^{-1}, \qquad (10.8a)$$

$$\beta = 1 + R^2. \qquad (10.8b)$$

Remembering that $m_1 > 0 > m_2$, we get (after considerable algebraic manipulation) Weil's solutions from Eqs. (10.4c,d) to (10.8a,b):

$$\frac{m_1}{\sigma_w} = \frac{1}{2}\left[\alpha S + \left(\alpha^2 S + \frac{4}{\beta}\right)^{1/2}\right], \qquad (10.9a)$$

$$\frac{m_2}{\sigma_w} = \frac{1}{2}\left[\alpha S - \left(\alpha^2 S + \frac{4}{\beta}\right)^{1/2}\right]. \qquad (10.9b)$$

With $R = 1$ so $\alpha = 1/2$ and $\beta = 2$, the above solutions are equivalent to those given by BB and by Luhar and Britter. Du et al. (1994) made use of Eqs. (10.4a–e) and assumed 1) $A_1 = 0.4$ and 2) the kurtosis $K = \langle w^4 \rangle/\sigma_w^4 = 3$, independent of the skewness $S = \langle w^3 \rangle/\sigma_w^3$.

Luhar et al. (1996) showed that the above three-moment closures with R constant do not result in Eq. (10.1) reducing to a Gaussian PDF as the skewness $S \to 0$, and that the four-moment closure of Du et al. (1994) does so only with the unrealistic assumption that $K = 3$, the value for Gaussian turbulence. Luhar et al. proposed that R in Eq. (10.6) not be a constant but a function of the skewness S:

$$R^{-1} = cS^{1/3}, \qquad (10.10)$$

where $c = 2/3$. Then Eqs. (10.4a–d), (10.6), and (10.10) give the following solution for the PDF of Eq. (10.1):

$$\frac{\sigma_1}{\sigma_w} = \left[\frac{A_2}{A_1(1 + R^{-2})^2} \right]^{1/2}, \qquad (10.11a)$$

$$\frac{\sigma_2}{\sigma_w} = \left[\frac{A_1}{A_2(1 + R^{-2})^2} \right]^{1/2}, \qquad (10.11b)$$

$$A_1 = \frac{1}{2} \left[1 - \left(\frac{r}{4 + r} \right)^{1/2} \right], \qquad (10.11c)$$

$$A_2 = 1 - A_1, \qquad (10.11d)$$

where

$$r = [(1 + R^{-2})^3 S^2]/[(3 + R^{-2})^2 R^2] \qquad (10.11e)$$

and the kurtosis is given by

$$K = (1 + r)(3 + 6R^{-2} + R^{-4})/(1 + R^{-2})^2. \quad (10.12)$$

The above closure reduces the bi-Gaussian PDF of Eq. (10.1) to a simple Gaussian PDF as $S \to 0$.

10.2 Pioneering applications

Baerentsen and Berkowicz (1984) used a pair of Langevin equations, one for updrafts and the other for downdrafts, each with Gaussian forcing so as to give an overall non-Gaussian PDF for the turbulence as described above. They chose to use the Legg and Raupach (1982) drift correction [they did not cite Wilson et al. (1983), who proposed the one-dimensional drift correction that meets the well-mixed condition]. They claimed good agreement with experimental data from water tanks.

De Baas et al. (1986) used the above procedure of BB to define the first three moments of a non-Gaussian forcing function for *one* Langevin equation. They did not discuss the problem of realizability. Their version of this equation for non-Gaussian turbulence is equivalent to that developed by Thomson (1984) section 3 whom they cited. The drift correction in this model reduces to that of Legg and Raupach in Gaussian turbulence. They

improved the CBL parameterization of BB and favorably compared their results with those of BB and data from experiments in water tanks, wind tunnels, and the field.

Sawford and Guest (1987) reviewed the work of BB and de Baas et al. and noted that the simpler one-equation model of the latter is the better of the two. However, they observed that the de Baas et al. model did not give an asymptotically uniform vertical concentration profile downwind, but gave 0.7 that value in thin layers at the top and bottom boundaries. They noted that Thomson (1984) and Sawford (1986) had shown that the Wilson et al. model meets the well-mixed condition, so they used that drift correction. They chose to use non-Gaussian forcing as did de Baas et al., but they were aware that a realizable non-Gaussian PDF has nonzero values for the third and all higher moments. They stated that it is impractical to specify all moments, and that only in rare circumstances are values known for more than the first three moments of the turbulence. They used the procedure of BB to define the first three moments of the non-Gaussian PDFs for both the initial turbulence field and the random forcing, realizing that they did not have an exact match between input and output. They obtained better results than BB and de Baas et al. These improvements came from a model with better turbulence parameterizations and more nearly correct boundary conditions at the top and bottom.

Flesch and Wilson (1992) extended the work of Baerentsen and Berkowicz to a joint PDF for the longitudinal and vertical velocities, u, w, in a model for turbulent diffusion within a plant canopy. They constructed their PDF for u, w as the sum of two joint Gaussian distributions. After some experimentation, they developed a system with nine unknowns: five equations for the zeroth through fourth moments of the vertical velocity w and four equations for the first through fourth moments of the longitudinal velocity u. They obtained good results with this model in homogeneous, slightly non-Gaussian turbulence. However, this model gave substantially poorer results in inhomogeneous and highly non-Gaussian turbulence than a model using Gaussian PDFs. This may have been a result of inaccurate formulation of the non-Gaussian PDF.

10.3 Model of Luhar–Britter and Sawford–Weil

The first definitive Langevin model with (realizable) Gaussian forcing that is consistent with a known approximation to non-Gaussian inhomogeneous turbulence was developed independently by Luhar and Britter (1989) and Sawford (Weil 1990). The BB procedure was used in each paper to define the first three moments of the non-Gaussian turbulence. All acknowledged that non-Gaussian forcing introduced problems, and they cited Thomson (1987) on realizability. The two models are equivalent, and one can be transformed into the other. Luhar and Britter presented the complete derivation; Weil gave the result and credited the derivation to Sawford (who provided it in a private communication in 1987). The complete derivation is lengthy and complex, so we will present an outline of what is presented by Luhar and Britter. We start with the Langevin equation,

$$dw = a(z,t)dt + b(z,t)dW(t), \qquad (10.13a)$$

and its companion,

$$dz = wdt. \qquad (10.13b)$$

The corresponding Fokker–Planck equation for *stationary* conditions is

$$w\frac{\partial P_{CBL}}{\partial z} + \frac{\partial}{\partial w}(aP_{CBL}) = \frac{1}{2}\frac{\partial^2}{\partial w^2}(b^2 P_{CBL}) \qquad (10.14a)$$

where

$$P_{CBL} = P_{CBL}(w,z) = P_{CBL}(w|z)P_{CBL}(z), \qquad (10.14b)$$

and the well-mixed condition (uniform tracer distribution in the vertical) is given by (Weil 1990)

$$P_{CBL}(z) = h^{-1}. \qquad (10.14c)$$

Here $P_{CBL}(w,z)$ is the joint unconditional Eulerian probability density of w,z in phase space, and $P_{CBL}(w|z)$ is the Eulerian probability density of w conditioned by z and defined by Eq. (10.1). We rewrite the above in two parts as

$$aP_{CBL} = \frac{\partial}{\partial w}\left(\frac{1}{2}b^2 P_{CBL}\right) + \phi, \qquad (10.15a)$$

where ϕ satisfies

$$\frac{\partial \phi}{\partial w} = -w\frac{\partial P_{CBL}}{\partial z} \qquad (10.15b)$$

and $\phi \to 0$ as $|w| \to \infty$. We simplify the PDF in Eq. (10.1) for convenience:

$$P_{CBL}(w|z) = A_1 P_1(w|z) + A_2 P_2(w|z). \qquad (10.16)$$

We use Eq. (3.24) in preference to Eq. (3.21) in defining b:

$$b = (C_0\varepsilon)^{1/2}. \qquad (10.17)$$

We combine Eqs. (10.12a) and (10.14) to get

$$a = \left(-\frac{C_0\varepsilon}{2}Q + \phi\right)P_{CBL}^{-1}, \qquad (10.18)$$

where

$$Q = \frac{A_1(w - m_1)P_1}{\sigma_1^2} + \frac{A_2(w - m_2)P_2}{\sigma_2^2}. \qquad (10.19)$$

The parameters $A_1, A_2, m_1, m_2, \sigma_1, \sigma_2$ are as given in Eqs. (10.1), (10.2a,b), and (10.11a–e). We then rewrite Eq. (10.13a) as

$$dw = \left[\frac{-(C_0\varepsilon/2)Q + \phi}{P_{CBL}}\right]dt + (C_0\varepsilon)^{1/2}dW(t). \qquad (10.20)$$

The random forcing is the Gaussian incremental Wiener process with zero mean and variance dt as defined in Eq. (4.7). The *general* solution for ϕ, *without any closure* for the six parameters introduced in Eq. (10.1) is given by Luhar et al. (1996):

$$\phi = \phi_1 + \phi_2, \qquad (10.21a)$$

where ϕ_1 and ϕ_2 are defined by the following with $i = 1,2$ (without summation):

$$\phi_i = \sigma_i^2 P_i \frac{\partial A_i}{\partial z} + \frac{A_i P_i}{2}\left(1 + \frac{w^2}{\sigma_i^2}\right)\frac{\partial \sigma_i^2}{\partial z}$$
$$\qquad - \frac{1}{2}\frac{\partial}{\partial z}(A_i m_i)\,\mathrm{erf}\left(\frac{w - m_i}{\sqrt{2}\sigma_i}\right)$$
$$\qquad + A_i P_i w\left(\frac{\partial m_i}{\partial z} - \frac{1}{2}\frac{m_i}{\sigma_i^2}\frac{\partial \sigma_i^2}{\partial z}\right). \qquad (10.21b)$$

If we use a *constant* value for R as in Eq. (10.6), the term following the error function is equal to zero and Eq. (10.21b) reduces to the solution given by Luhar and Britter (1989) and Weil (1990). The quantity $-(C_0\varepsilon/2)(Q/P_{CBL})$ is the "fading memory" component of the deterministic term and the "drift correction" is ϕ/P_{CBL}. As we have noted in section 7.2, the use of Eq. (10.14c) ensures that the well-mixed condition is one of the possible solutions. The above solution is *stationary* unlike those in sections 7.2 and 8.4–8.6.

Luhar and Sawford (1995) extended the above model to two dimensions, incorporating streamwise diffusion and gradients of flow properties in both the streamwise and vertical dimensions. They applied this model to the coastal fumigation problem.

10.4 Four-moment model of Du, Wilson, and Yee

Du et al. (1994) used principles of information theory to formulate velocity PDFs for the convective boundary layer that 1) reflect the available information about the turbulence statistics and 2) quantitatively maximize the "missing information." It follows from this formalism that there must be an *even* number of moment constraints used to define the PDF. Therefore, they used the first four moments of the velocity (i.e., the mean, variance, skewness, and kurtosis). They calculated two PDFs; one was a "correct" PDF based on information theory and the other was a four-moment approximation based on Eqs. (10.4a–e). For the latter, their solution for ϕ_i can be obtained from Eqs. (10.21a,b). They found that the four-moment approximation (with $A_1 = 0.4$ and $K = 3$) was a better match to the theoretically "correct" PDF than the three-moment PDF with R in Eq. (10.6) constant. They

formulated two stochastic Lagrangian models, one based on the "correct" PDF and the other on the four-moment approximation. In each case, there was better agreement with convection tank data than with the original bi-Gaussian model of Luhar–Britter and Sawford–Weil. The "correct" model required significantly more computer time, but did not yield significantly better performance than the four-moment approximation.

10.5 Closure of Luhar, Hibberd, and Hurley

We have already discussed some features of the closure proposed by Luhar et al. (1996) and presented in Eqs. (10.10)–(10.12) and Eqs. (10.21a,b). Luhar, Hibberd, and Hurley fitted CBL velocity statistics from a saline water tank test facility to obtain an empirical velocity PDF based on Eq. (10.1). They used this PDF in comparisons of dispersion model performance with the PDF closures of 1) Baerentsen and Berkowicz (1984) and Luhar and Britter (1989) with $R = 1$ in Eq. (10.6), 2) Weil (1990) with $R = 3/2$, 3) Du et al. (1994) with $A = 0.4$ and $K = 3$ in Eqs. (10.4a–e), and 4) their PDF based on Eqs. (10.10)–(10.12). They concluded that their closure and that of Weil gave the best results. They recommended their closure, in part because the bi-Gaussian PDF based on this closure collapses to a simple Gaussian as the skewness goes to zero.

Chapter 11

The Boundary Condition Problem

We have not discussed one problem that must be faced in using the stochastic Lagrangian models of turbulent diffusion that we studied in the preceding chapters: How should we treat the trajectories of "marked particles" near the boundaries of the turbulent atmospheric boundary layer? The following discussion is in the context of two-dimensional simulations with vertical (z) and longitudinal (x) coordinates. We are concerned with the boundaries at the bottom and the top. The lower boundary is the bottom of the computational domain that generally corresponds to the land or water surface. This boundary can act as a perfect reflector, a sink (deposition), or a source (evaporation, resuspension, etc.). There can be two upper boundaries: that of the computational domain and, at a lower level, that of the turbulent atmospheric boundary layer. In the latter case, there can be entrainment into or out of the boundary layer.

11.1 Historical review

The boundary condition problem is not mentioned in many papers in the literature, it is a minor topic in some, and it is discussed in depth in only a few. Boundary conditions for the trajectories of marked particles are described in the following papers, but there is not discussion of any problems associated with these conditions. Ley (1982) and Ley and Thomson (1983) assumed perfect reflection (to be defined in section 11.2) at $z = z_0$, the surface roughness height, in their simulations of diffusion in the surface layer. Thomson (1984) assumed perfect reflection at the bottom $(z = 0)$ and top $(z = h)$ of turbulent flow in a channel. Thomson (1987) section 5.2 assumed perfect reflection at $z = 0$, the bottom of a free convection layer. [We did not include Thomson (1987) in our review in Chapter 10 because, as noted by Luhar and Britter (1989), that model of Thomson's is valid for only the lower part of the CBL where the skewness of the turbulence is very small.] Baerentsen and Berkowicz (1984) and de Baas et al. (1986) assumed perfect reflection at both the bottom $(z = 0)$ and top $(z = h)$ of the CBL.

We cited Boughton et al. (1987) in section 6.3 on the Markov or diffusion limit of the Langevin equation, the random displacement model. They made the comment that, in earlier experience with stochastic models of atmospheric diffusion, considerable difficulties were experienced in correctly simulating other than perfect reflection at the boundaries, but they did not cite any specific cases. They used a transition probability density in their boundary condition for particle deposition on the ground (a principal contribution of their paper), and a later note (1989) clarified some questions that had been raised. Perfectly reflecting boundary conditions can cause problems. Two examples were mentioned by van Dop et al. (1985). First, numerical accuracy requires that $\Delta t \ll \tau_L$ where Δt is the incremental time step and τ_L is the Lagrangian decorrelation timescale (in inhomogeneous turbulence, the local decorrelation timescale). This condition may be violated close to a boundary if τ_L is a function of position [for example, $\tau_L = \tau_L(z)$] as in Ley's (1982) simulation cited above. Second, van Dop et al. observed that the vanishing of τ_L near a boundary (as in Ley) raises the question of whether or not particles can reach such a boundary. Under some conditions, the answer is "no." They cautioned that care should therefore be used in predicting surface concentrations. They also commented that it is likely that finite-difference calculations of surface concentrations are dependent on grid size.

Sawford and Guest (1987) were specific in discussing problems associated with the perfectly reflecting boundaries at the top and bottom in de Baas et al. (1986). They observed that the model of de Baas et al. did not preserve the well-mixed condition far downwind, but gave concentrations about 0.7 times the homogeneous value in two thin layers along the top and bottom boundaries. They noted that the reflection condition requires that the skewness of the turbulence be zero at the boundary, and that the model of de Baas et al. does not meet this condition. They demonstrated that a modified de Baas et al. model with zero skewness gave much better results at the top (and worse results at the bottom because of the turbulence inhomogeneity near that boundary). Their model gave better results in inhomogeneous turbulence, and they used parameterizations for the vertical velocity variance and skewness that vanished at the reflecting boundaries. Consequently, these boundaries could not be reached so they presented near-surface concentrations, not surface concentrations, as results. They also noted that the perfect reflection condition at the top of the CBL is not strictly correct.

Luhar and Britter (1989) used the same boundary conditions and the same velocity variance and skewness parameterizations as Sawford and Guest. Their only comment about problems at the boundaries was that the use of large values of Δt did not give the proper behavior of particles near boundaries. They used a very large time step everywhere. In some places, Δt exceeded the local timescale τ_L, so they used an ad hoc procedure to overcome the resultant difficulties.

Wilson et al. (1989) considered the case of a partially absorbing lower boundary for stochastic Lagrangian models. They gave a reflection probability (used in conjunction with conditional perfect reflection) that corresponds to a specified deposition velocity.

Weil's (1990) contribution to this problem is presented for homogeneous skewed turbulence in section 3.b and appendix B of his paper. He modified the reflection boundary condition to account for the asymmetry of the probability density function for the vertical velocity fluctuations in the CBL.

11.2 Study by Wilson and Flesch

Wilson and Flesch (1993) is a pioneering paper [see Wilson and Flesch (1992) for a preview presented at a symposium] in which they explicitly addressed the boundary condition problem in the context of a three-component system:

1) A Langevin equation model that meets the well-mixed criterion of Thomson (1987),

2) Algorithms for discrete-time (Δt) calculations throughout the flow (including near boundaries), and

3) Parameterizations of turbulence statistics [e.g., $\sigma_w^2 = \sigma_w^2(z)$ and $\tau_L = \tau_L(z)$] for use as inputs.

They demonstrated that a combination of components 1–3 could result in violation of the well-mixed criterion if either component 2 or component 3 is not properly formulated. Specifically, they found

that the perfect reflection algorithm is exactly consistent with the well-mixed constraint (their term which they abbreviated as "wmc") in homogeneous Gaussian turbulence, but that no reflection scheme can satisfy this criterion if the probability distribution of the velocity w normal to the boundary is skewed ($\langle w^3 \rangle \neq 0$) or inhomogeneous ($\partial \sigma_w^2 / \partial z \neq 0$) where $z \to 0$. They made two interesting statements about the wmc:

> "...passive tracer particles [that are] initially well mixed in the flow ... *must remain well mixed, with respect to both position and velocity.*"

> "...there is no dynamical content to the wmc. The wmc is rather a statement with respect to entropy, ... prohibiting the spurious evolution of order from disorder."

The emphasis in the first statement is in the original. We summarize their results in the following paragraphs, which approximately parallel their sections beginning with section 2.

There is normally a layer of the atmosphere adjacent to the land or water surface in which the turbulence statistics are unknown. They called it the "unresolved basal layer" or UBL. They concluded that the "true" statistics for the UBL will never be known and that there is no barrier to pragmatism in their specification. They were pragmatic in modifying parameterizations of turbulence statistics in the UBL to ensure that the well-mixed criterion is met.

It is possible to specify the vertical profiles of the velocity statistics (σ_w^2, τ_L) and the time step (Δt) in such a way that the boundary (e.g., land or water surface) is never reached (i.e., is naturally unobtainable). In such a case, the above system of components 1–3 meets the well-mixed criterion, but at the expense of increased computational effort. They found that it is more efficient to use a properly defined trajectory reflection. They also experimented with "artifical unattainability" and found that it inevitably violated the well-mixed criterion.

Wilson and Flesch used the companion to the Chapman–Kolmogorov equation in formulating the well-mixed criterion, whereas Thomson (1987) used the Fokker–Planck equation. We begin with their Eq. (2), which we write for z, w phase space in the same form as our Eq. (5.5) for the state y:

$$P_1(z_2, w_2, t_2) =$$
$$\int P_{1|1}(z_2, w_2, t_2 | z_1, w_1, t_1) P_1(z_1, w_1, t_1) dz_1 dw_1.$$

(11.1)

Here $P_1(z_1, w_1, t_1)$ is the joint probability density of velocity w and position z at time t_1, $P_1(z_2, w_2, t_2)$ is the joint probability density of z, w at time t_2, $P_{1|1}(z_2, w_2, t_2 | z_1, w_1, t_1)$ is the conditional or transition joint probability density of location z_2 and velocity w_2 at time t_2, given that it was at z_1 and w_1 at t_1, and $t_1 < t_2 < t_3$, etc. This is the first condition for a Markov process (but it can also apply to other processes). It states that the velocity w_2 at position z_2 at time t_2 can have its origin in any of the probable states z_1, w_1 at t_1. We rewrite our Eq. (5.6), the Chapman–Kolmogorov equation, for z, w phase space:

$$P_{1|1}(z_3, w_3, t_3 | z_1, w_1, t_1) =$$
$$\int P_{1|1}(z_3, w_3, t_3 | z_2, w_2, t_2)$$
$$\times P_{1|1}(z_2, w_2, t_2 | z_1, w_1, t_1) dz_2 dw_2.$$
$$(11.2)$$

This is the second condition for a Markov process. It statistically decouples the process values z_3, w_3 at time t_3 (the future) from the values z_1, w_1 at time t_1 (the past) because, for a given past, there is a wide range of possible values for z_2, w_2 at time t_2 (the present). As we stated in section 5.3, the combination of the unnamed Eq. (11.1) and the Chapman–Kolmogorov equation, Eq. (11.2), defines a Markov process, a process for which there is neither a unique past nor a unique future for a given present. We follow Wilson and Flesch in restricting this study to constant fluid density and stationary flows, so the Eulerian PDF, $P_E(z, w, t)$, is independent of time. We identify P_E with P_1 in Eq. (11.1) with the result that

$$P_E(z_2, w_2, t_2) =$$
$$\int P_{1|1}(z_2, w_2, t_2 | z_1, w_1, t_1) P_E(z_1, w_1, t_1) dz_1 dw_1,$$
$$(11.3)$$

which is equivalent to their Eq. (3). Any stochastic Lagrangian model (i.e., transition probability $P_{1|1}$) satisfies the well-mixed criterion if it has the property defined by Eq. (11.3). That is, *it maps any well-mixed state to a subsequent state that is also well-mixed.* The transition probability $P_{1|1}$ is defined by the three-component system described above (the Langevin equation model, algorithms for finite time steps Δt, and parameterizations of turbulence statistics). If a system fails the well-mixed criterion, it *must* fail on the *first* time step, so we let

$$t_0 = t \ and \ t_1 = t + \Delta t. \qquad (11.4)$$

Here and in the following, our transition is from state 0 to state 1 instead of from 1 to 2 as given in Eqs. (11.1)–(11.3) in order to be consistent with the notation of Wilson and Flesch (1993).

Wilson and Flesch considered the possibility of guidelines for a reflection scheme that are more restrictive than those implicit in Eq. (11.3) and the subsequent discussion. They noted that reflection schemes break the equivalence between a model and real time (in terms of the time since the release of the particle). That is, tracer particles do not "bounce" off the ground or water surface as in a reflection algorithm. They noted that such an algorithm must do more than ensure that (in the absence of deposition or resuspension) there is zero flux of both the fluid and the tracer through the boundary; in addition, it must not distort the velocity PDF in the vicinity of $z = 0$. That is, the algorithm must satisfy Eq. (11.3). If not, it violates the well-mixed condition. They concluded that meeting the well-mixed criterion as defined by Eq. (11.3) is sufficient.

Wilson and Flesch presented the procedure for defining the transition probability density $P_{1|1}$ in Eq. (11.3) for a simple model of one-dimensional (vertical) diffusion in a system bounded at the top and bottom. They used finite-difference versions of Eqs. (7.2a,b) in which the coefficient b of the incremental Wiener process, $\Delta W(t)$, is constant. Their reflection algorithm is operational if a particle moves from an allowed state (z_0, w_0, t_0) to a forbidden state (z^*, w^*, t_1) on the other side of a boundary. They assumed perfect reflection so $w_1 = -w^*$ and z_1 is at an appropriate distance below the top boundary or above the bottom boundary. They defined the transition probability density $P_{1|1}$ for a single time step as the product of independent transitions for position z and velocity w; that is, $P_{1|1} = P_z P_w$. P_z is a function of the algorithm for position change, and P_w is a Gaussian function of the algorithm for velocity change. They combined the algorithm-based expressions for P_z and P_w to obtain an equation for $P_{1|1}$. They substituted Eq. (7.6) for P_E and the relation $P_{1|1} = P_z P_w$ into Eq. (11.3), integrated the right side of the result over all values of w_0 and z_0, and found that the well-mixed criterion, $P_E(z, w, \Delta t) = P_E(z_0, w_0, 0)$, is met if

$$b^2 = \frac{\sigma_w^2}{\Delta t} \left[1 - \left(1 - \frac{\Delta t}{\tau_L} \right)^2 \right]. \qquad (11.5)$$

This equation reduces to the familiar $b^2 = 2\sigma_w^2/\tau_L$ as $\Delta t \to 0$ [see Eqs. (3.24) and (3.25)]. Wilson and Flesch admitted that their analysis proving the above is laborious. In order not to interrupt this narrative with a lot of mathematics, we present the procedure that leads to Eq. (11.5) in the appendix to this chapter.

Wilson and Flesch empirically confirmed the above procedure in numerical simulations. They also demonstrated that a two-dimensional (vertical and longitudinal) turbulence model by Thomson (1987) requires reflection of *both* components of the particle velocity, even though this requirement for the longitudinal velocity has no "real world" significance. They numerically integrated an expanded form of Eq. (11.3) using a constant velocity variance σ_w^2 and a timescale τ_L that increased linearly with height, and found that the well-mixed criterion was met if the ratio $\Delta t/t_L$ was kept very small. They concluded that their reflection algorithm can be applied in Gaussian turbulence with height-dependent velocity variance if the gradient of the variance $\partial \sigma_w^2/\partial z$ vanishes within an appropriate scale length above the surface. The existence of the "unresolved basal layer" (mentioned in the beginning of this section) gives one the freedom to use this artifice.

They studied reflection algorithms for skewed turbulence and concluded in a lengthy section that no reflection scheme examined preserved the well-mixed criterion if the turbulence in the vicinity of the reflecting surface is skewed. They used numerical integration of a variation of Eq. (11.3) written for homogeneous skewed turbulence to compare perfect reflection with a proposal by Weil (1990). They found that both methods failed the well-mixed criterion, but that Weil's proposal did a better job of preserving the skewness. They suggested that the skewness, like the velocity variance, be parameterized so that it vanishes near a boundary. [More recently, Thomson and Montgomery (1994), using numerical simulations, found that an algorithm similar to, but more elaborate than Weil's method, performed better than Weil's algorithm, but both were satisfactory with sufficiently small Δt. This paper is recommended for study.]

Wilson and Flesch tested the models of Luhar and Britter (1989) and Weil (1990) for the convective boundary layer. The boundaries are naturally unattainable in the Luhar and Britter model, and Weil's model includes a reflection algorithm. They concluded that both models can be implemented so as to satisfactorily approximate, if not meet, the well-mixed criterion. There are many other details in Wilson and Flesch (1993) not given here that are worthy of study.

11.3 The random displacement model

Durbin (1983) considered three types of boundary conditions for the random displacement model on pp. 17–23: nonattainable, reflecting, and absorbing. He noted that "nonattainable" boundaries can be reached in finite-difference computations, and he suggested an algorithm to correct this condition. He also presented algorithms for reflecting and absorbing boundaries. He made reference to the boundary condition discussion in chapter 5 of Gihman and Skorohod (1972).

Boughton et al. (1987, 1989) discussed boundary conditions for deposition as a function of a transition probability density for absorption at the surface. This transition probability density, attributed to Monin (1959), is a function of the settling velocity, the deposition velocity, and the diffusivity at the surface. Their 1989 note calls attention to a typographical error in one of the equations in Monin's paper.

11.4 Appendix: A reflection algorithm

Wilson and Flesch (1993) demonstrated a procedure for finding the transition probability $P_{1|1}$ for a simple model of one-dimensional (vertical) diffusion in a system bounded at both the top and bottom where $z = T$ and $z = B$, respectively. We further simplify our demonstration by considering reflection at only the bottom where $z = 0$ (as in Wilson and Flesch 1992). We begin with the model given by Eqs. (7.2a,b), but written in finite-difference form. We calculate the incremental displacement,

$$\Delta z = z_1 - z_0 = w_0 \Delta t, \qquad (11.6a)$$

before incrementing the velocity,

$$\Delta w = w_1 - w_0 = a(z_0, w_0, t)\Delta t + b\Delta W(t). \quad (11.6b)$$

Note that we follow Wilson and Flesch (1992, 1993) by assuming the coefficient b of the incremental Wiener process $\Delta W(t)$ to be constant. We continue with the description of their reflection algorithm. Suppose that a particle goes from an allowed state (z_0, w_0, t) above the boundary to a forbidden state $(z^*, w^*, t + \Delta t)$ below the boundary where $z = 0$.

This will occur only if $w_0 > -z_0/\Delta t$. If this happens, we assume perfect reflection so the new particle position $z_1(w_0, t + \Delta t)$ is above the boundary at $z_1 = z^*$ and the direction of the velocity is reversed so $w_1 = -w^*$. We introduce a unit step function $C(z, w)$ that takes the following values:

$$C(z, w) = 0 \text{ if } w_0 \geq -z_0/\Delta t \qquad (11.7a)$$

so there is no reflection and

$$C(z, w) = 1 \text{ if } w_0 < -z_0/\Delta t \qquad (11.7b)$$

so reflection occurs. This algorithm is incorporated into Eqs. (11.6a,b) as follows:

$$\Delta z = z_1 - z_0 = w_0\Delta t(1 - 2C) - 2Cz_0, \quad (11.8a)$$

$$\Delta w = w_1 - w_0 =$$
$$a(z_0, w_0, t)\Delta t(1 - 2C) - 2C_0w_0 + b(1 - 2C)\Delta W(t).$$

$$(11.8b)$$

Equations (11.8a,b) are a combination of a stochastic Lagrangian model and a reflection algorithm together with a parameterization for $a(z_0, w_0, t)$, that form a complete three-component system. Equations (11.7a,b) and (11.8a) gives two possible locations for the new particle position z_1: the end point of a direct trajectory if Eq. (11.7a) applies,

$$z_1 = z_0 + w_0\Delta t, \qquad (11.9a)$$

and that of a reflected trajectory if Eq. (11.7b) is applicable,

$$z_1 = -z_0 - w_0\Delta t. \qquad (11.9b)$$

Similarly, from the combinations of Eqs. (11.7a,b) and (11.8b), we get

$$w_1 = w_0 + a(z_0, w_0, t)\Delta t + b\Delta W(t) \qquad (11.10a)$$

without reflection, and

$$w_1 = -w_0 - a(z_0, w_0, t)\Delta t - b\Delta W(t), \quad (11.10b)$$

where we get a change in sign (i.e., reversed direction) with reflection. Note that the random process $\Delta W(t)$ is *not* involved in the move from z_0 to z_1, but *only* in the change in velocity from w_0 to w_1.

The transition probability density $P_{1|1}$ for a single time step Δt is the product of independent transitions for position z and velocity w:

$$P_{1|1} = P_zP_w. \qquad (11.11)$$

The transition probability density P_z for changes in position from z_0 to z_1 is a function of the algorithm for position change, Eq. (11.8a):

$$P_z(z, t + \Delta t|z_0, w_0, t) =$$
$$\delta[(z - z_0), (w_0\Delta t(1 - 2C) - 2Cz_0)],$$

$$(11.12)$$

where δ is the Dirac delta function. That is, P_z is equal to zero unless $z - z_0 = \Delta z$ as defined by Eq. (11.8a), in which case it is equal to unity. The random forcing $b\Delta W(t)$ is Gaussian so the transition probability density for changes in velocity is also Gaussian. From Eqs. (4.7a,b) and (11.8b), the mean of the velocity transition from w_0 to w_1 is

$$\langle w - w_0 \rangle = a(z_0, w_0, t)\Delta t(1 - 2C) - 2Cz_0, \quad (11.13a)$$

and its variance is

$$\langle (w - w_0)^2 \rangle = b^2\Delta t. \qquad (11.13b)$$

Then the transition probability density P_w for the change in velocity from w_0 to w_1 is

$$P_w(w, t + \Delta t|z_0, w_0, t) =$$
$$\frac{1}{(2\pi)^{1/2}b(\Delta t)^{1/2}} \exp\left[-\frac{(w - w_0 - \langle w - w_0\rangle)^2}{2\langle(w - w_0)^2\rangle}\right],$$

$$(11.14)$$

where $\langle w - w_0 \rangle$ and $\langle (w - w_0)^2 \rangle$ are defined by Eqs. (11.13a–b), respectively. Finally, we get $P_{1|1}$ from Eqs. (11.11)–(11.14):

$$P_{1|1}(z, w, \Delta t|z_0, w_0, 0) =$$
$$\frac{\delta[(z - z_0), (w_0\Delta t(1 - 2C) - 2Cz_0)]}{(2\pi)^{1/2}b(\Delta t)^{1/2}}$$
$$\times \exp\left[-\frac{(w - [w_0 + a(z_0, w_0, 0)\Delta t](1 - 2C))^2}{2b^2\Delta t}\right],$$

$$(11.15)$$

where C is determined by Eqs. (11.7a,b). Equation (11.15) for our discrete-time Markov process differs at $O(\Delta t)^2$ from the analogous Eq. (7.6) that we used in deriving Eq. (7.18) from Eq. (7.2a).

We modify Eq. (7.18) for one-dimensional (vertical) diffusion that meets the well-mixed criterion by assuming assume homogeneous turbulence with constant velocity variance σ_w^2 and timescale τ_L, replace the coefficient of the incremental Wiener process with b, and write the simplified result as follows:

$$w - w_0 = -\frac{w_o}{\tau_L}\Delta t + b\Delta W(t). \qquad (11.16)$$

Here $-w_0/\tau_L$ replaces the quantity $a(z_0, w_0, t)$ in Eq. (11.6b) and subsequent equations. We incorporate the reflection algorithm, Eqs. (11.7a–b), to get the following:

$$w - w_0 = -\frac{w_0}{\tau_L}\Delta t(1-2C) - 2Cw_0 + b\Delta W(t)(1-2C). \qquad (11.17)$$

We combine Eq. (11.15) for the transition probability density with Eq. (11.17) for the change in velocity; first without reflection [Eq. (11.7a)] and then with reflection [Eq. (11.7b)]. The transition probability density without reflection ($w_0 \geq -z_0/\Delta t$) is

$$P_{1|1}(z, w, \Delta t | z_0, w_0, 0) =$$
$$\frac{(1/\Delta t)\delta[w_0, (z - z_0)/\Delta t]}{(2\pi)^{1/2}b(\Delta t)^{1/2}}$$
$$\times \exp\left(-\frac{[w - w_0(1 - \Delta t/\tau_L)]^2}{2b^2\Delta t}\right),$$

$$(11.18a)$$

where we have rewritten the transition probability for position in terms of a Dirac delta function of the velocity w_0. The transition probability density with reflection ($w_0 < -z_0/\Delta t$) is

$$P_{1|1}(z, w, \Delta t | z_0, w_0, 0) =$$
$$\frac{(1/\Delta t)\delta[w_0, -(z + z_0)/\Delta t]}{(2\pi)^{1/2}b(\Delta t)^{1/2}}$$
$$\times \exp\left(-\frac{[w + w_0(1 - \Delta t/\tau_L)]^2}{2b^2\Delta t}\right),$$

$$(11.18b)$$

where we have rewritten the Dirac delta function in a similar manner. We introduce the well-mixed criterion defined by Eq. (7.3) for the Gaussian distribution of the velocity of both the fluid and the tracer,

$$P_E(z, w, t) = (2\pi)^{-1/2}\sigma_w^{-1}\exp\left[-\frac{1}{2}\left(\frac{w}{\sigma_w}\right)^2\right],$$

$$(11.19)$$

where σ_w is constant in stationary homogeneous turbulence. We rewrite Eq. (11.3) in terms of Eqs. (11.18a,b) and (11.19):

$$P_E(z, w, \Delta t) = \int_{z_0=0}^{\infty}\left[\int_{w_0=-\infty}^{-z_0/\Delta t} P_{1|1}(z, w, \Delta t | z_0, w_0, 0)P_E(z_0, w_0, 0)dw_0\right]dz_0$$
$$+ \int_{z_0=0}^{\infty}\left[\int_{w_0=-z_0/\Delta t}^{\infty} P_{1|1}(z, w, \Delta t | z_0, w_0, 0)P_E(z_0, w_0, 0)dw_0\right]dz_0,$$

$$(11.20)$$

where the first integral on the right is for reflection and the second is for no reflection. According to Wilson and Flesch, this is "readily integrated" to confirm that the well-mixed criterion, $P_E(z, w, \Delta t) = P_E(z_0, w_0, 0)$, is met, provided that

$$b^2 = \frac{\sigma_w^2}{\Delta t}\left[1 - \left(1 - \frac{\Delta t}{\tau_L}\right)^2\right], \qquad (11.21)$$

which reduces to the familiar $b^2 = 2\sigma_w^2/\tau_L$ as $\Delta t \to 0$ [see Eqs. (3.24) and (3.25)]. They admitted that their analysis proving that the above exactly satisfies the well-mixed criterion is laborious. We outline their procedure (J. Wilson 1994, personal communication) in the following.

We begin by combining Eqs. (11.18a,b)–(11.20). After integrating with respect to dw_0, we get

$$P_E(z, w, \Delta t) = \frac{1}{2\pi b \sigma_w \Delta t^{3/2}} \int_{z_0=0}^{\infty} \exp\left[-\frac{[w - (z + z_0)(\eta/\Delta t)]^2}{2b^2\Delta t} - \frac{(z + z_0)^2}{2\sigma_w^2\Delta t^2}\right] dz_0$$
$$+ \int_{z_0=0}^{\infty} \frac{1}{2\pi b \sigma_w \Delta t^{3/2}} \int_{z_0=0}^{\infty} \exp\left[-\frac{[w - (z - z_0)(\eta/\Delta t)]^2}{2b^2\Delta t} - \frac{(z - z_0)^2}{2\sigma_w^2\Delta t^2}\right] dz_0,$$

$$(11.22)$$

where

$$\eta = 1 - \Delta t/\tau_L. \tag{11.23}$$

The first integral in Eq. (11.22) is for reflection and the second is for no reflection. Our next step involves much algebraic manipulation to separate the terms with the zeroth, first, and second powers of z_0. In the process, we let

$$\xi = \frac{\eta^2}{b^2\Delta t^3} + \frac{1}{\sigma_w^2\Delta t^2}. \tag{11.24}$$

Then Eq. (11.22) takes the form

$$P_E(z, w, \Delta t) = \frac{\exp[-(w^2/2b^2\Delta t) + (wz\eta/b^2\Delta t^2) - (z^2\xi/2)]}{2\pi\sigma_w\Delta t^{3/2}}$$
$$\times \left[\int_{z_0=0}^{\infty} \exp\left[z_0\left(\frac{w\eta}{b^2\Delta t^2} - z\xi\right) - \frac{z_0^2\xi}{2}\right] dz_0 + \int_{z_0=0}^{\infty} \exp\left[-z_0\left(\frac{w\eta}{b^2\Delta t^2} - z\xi\right) - \frac{z_0^2\xi}{2}\right] dz_0\right].$$

$$(11.25)$$

From Eq. (3.332-2), p. 355 in Gradshteyn and Ryzhik (1994):

$$\int_0^{\infty} \exp\left(-\frac{x^2}{4\beta} - \gamma x\right) dx = \sqrt{\pi\beta} \exp(\beta\gamma^2)[1 - \phi(\gamma\sqrt{\beta})], \tag{11.26}$$

where ϕ is the error function. We note that the error function is an odd function (Spanier and Oldham 1987) so $\phi(\gamma\sqrt{\beta}) = -\phi(-\gamma\sqrt{\beta})$. Let

$$\beta = 1/2\xi \tag{11.27a}$$

and

$$\gamma = z\xi - w\eta/b^2\Delta t^2. \tag{11.27b}$$

Then the solution of Eq. (11.25) is

$$P_E(z, w, \Delta t) = \frac{\exp[-(w^2/2b^2\Delta t) + (wz\eta/b^2\Delta t^2) - (z^2\xi/2) + \beta\gamma^2]}{\sqrt{\pi}\beta^{-1/2} b \sigma_w \Delta t^{3/2}}. \tag{11.28}$$

We continue by substituting Eqs. (11.23), (11.24), and (11.27a,b) into Eq. (11.28) to get the simplified result,

$$P_E(z, w, \Delta t) = \frac{\exp\left[-w^2/2(\sigma_w^2(1 - \Delta t/\tau_L)^2 + b^2\Delta t)\right]}{\sqrt{2\pi}[\sigma_w^2(1 - \Delta t/\tau_L)^2 - b^2\Delta t]^{1/2}}. \tag{11.29}$$

Then, if Eq. (11.21) is applied, Eq. (11.29) reduces to Eq. (11.19) and the well-mixed criterion is met.

Chapter 12

Parameterization of Turbulence Statistics for Model Inputs

The principal inputs to the above stochastic Lagrangian models of turbulent diffusion [variations of the Langevin equation model (LEM) or the random displacement model (RDM)] are 1) the variance of the turbulent velocity fluctuations (e.g., σ_w^2) and 2) either the local decorrelation timescale τ_L or the rate of dissipation of the turbulence kinetic energy ε. The inputs σ_w^2 and τ_L or ε are sufficient for diffusion in one dimension, the vertical. In more complex cases (e.g., the convective boundary layer or two- or three-dimensional dispersion), the skewness $\langle w^3 \rangle / \sigma_w^3$ and kurtosis $\langle w^4 \rangle / \sigma_w^4$ of the vertical velocity fluctuations, the horizontal velocity variances σ_u^2 and σ_v^2, and the shear stresses (e.g., τ_{uw}) may be required. These inputs are generally assumed to be 1) strong functions of the vertical coordinate and the atmospheric stability, and 2) relatively weak functions of time and the horizontal coordinates, consistent with our simplifications of the LEM for two and three dimensions in chapter 9.

We are ignorant of much of the complex physics in the turbulent atmospheric boundary layer. We are limited by the closure problem of turbulence theory and the scarcity of quality data from physical measurements and numerical simulations. As a consequence, we are forced to use similarity theory in combination with empirical parameterizations in the following.

We use the scaling parameters and diagram of Holtslag and Nieuwstadt (1986), shown in Fig. 12.1, for dividing the idealized boundary layer of the atmosphere into distinct scaling regimes. These are functions of the dimensionless height z/h and the dimensionless stability parameter h/L where z is the vertical coordinate, h is the height of the turbulent boundary layer, and L is the Monin–Obukhov scaling length. We assume stationary, horizontally homogeneous turbulence statistics, consistent with chapter 9. After reviewing basic scaling parameters, we examine published parameterizations of the above turbulence statistics that have been, or could be, used as model inputs.

In general, the published parameterizations for adjacent scaling regimes in the Holtslag–Nieuwstadt

diagram *do not* give *continuous* values for the model inputs [e.g., $\sigma_w^2(z/h, h/L)$] across "fuzzy" boundaries between these regimes (e.g., the unstable free convection layer and the mixed layer as shown in Fig. 12.1. As a consequence, there is a need for parameterizations of inputs that are *continuous* functions of z/h and h/L over the entire parameter space illustrated in Fig. 12.1. We do not attempt to fill this need because most of the required work has not been done. We present the following examples only as approaches that *could* be used. We emphasize that *there is no unique way* to provide inputs for the LEM and RDM.

12.1 Basic scaling parameters

The scaling velocity for turbulence in stable and near-neutral surface boundary layers is defined by the equation

$$-\langle uw \rangle_0 = u_*^2, \qquad (12.1)$$

where u and w are, respectively, the longitudinal and vertical turbulent velocity fluctuations, $\langle uw \rangle_0$ is the Reynolds-averaged turbulent kinematic shear stress at the surface, and u_* is the friction velocity. A companion relation is

$$-\langle w\theta \rangle_0 = u_* \theta_*, \qquad (12.2)$$

where θ is the turbulent fluctuation of the temperature, and $\langle w\theta \rangle_0$ is the Reynolds-averaged temperature flux for turbulent heat transfer. A combination of the buoyancy parameter g/T (where g is the acceleration of gravity and T is the absolute temperature), u_*, and θ_* defines the Monin–Obukhov length scale:

$$L = u_*^2 T / kg\theta_* = -u_*^3 T / kg\langle w\theta \rangle_0, \qquad (12.3)$$

where k is the von Kármán constant ($k \approx 0.4$). The length scale is more convenient to use than the quantity g/T; furthermore $L > 0$ in stable conditions and $L < 0$ in unstable conditions.

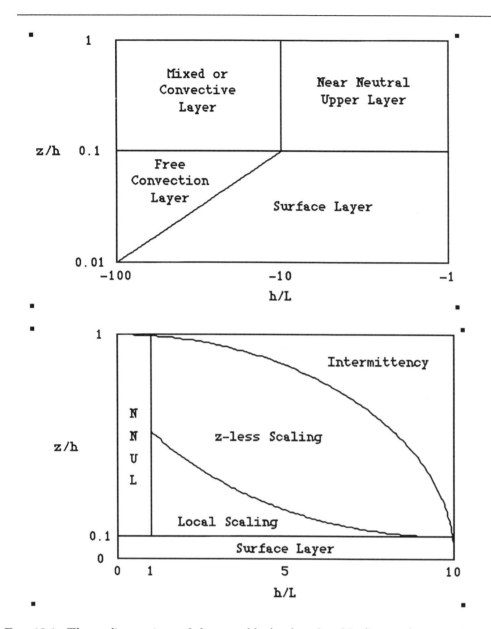

FIG. 12.1. The scaling regions of the unstable (top) and stable (bottom) atmospheric boundary layers, shown as functions of the dimensionless height z/h and the stability parameter h/L. A detailed discussion can be found in Holtslag and Nieuwstadt (1986). Reprinted by permission of Kluwer Academic Publishers and A.A.M. Holtslag.

The product of g/T, u_*, and θ_* is

$$\varepsilon_B = -gu_*\theta_*/T = \langle w\theta \rangle_0 g/T, \qquad (12.4)$$

where ε_B is the surface rate of buoyancy production ($\langle w\theta \rangle_0$ is positive) or destruction ($\langle w\theta \rangle_0$ is negative) of turbulence kinetic energy. The shear-stress production rate of turbulence kinetic energy is

$$\varepsilon_S = u_*^3/kz, \qquad (12.5)$$

where z is the height above the surface. The minimum value of z is z_0, the surface roughness length. The ratio of the rates of buoyant to shear-stress production of turbulence kinetic energy is [Eqs. (12.3)–(12.5)]

$$\varepsilon_B/\varepsilon_S = -z/L. \qquad (12.6)$$

This shows that the effect of buoyancy approaches zero as $|z/L| \to 0$, and the effect of shear stress becomes negligible for sufficiently large values of $|z/L|$.

In stable boundary layers ($L > 0$), Eqs. (12.3)–(12.6) indicate that the destruction of turbulence kinetic energy by negative buoyancy will limit the height at which turbulence can be maintained. On the other hand for unstable boundary layers ($L < 0$), these equations imply that u_* is not an appropriate scaling velocity for turbulence if the effect of shear stress is negligible. A more apropriate scaling velocity is the convective velocity scale, which is defined as

$$\varepsilon_B h = \langle w\theta \rangle_0 gh/T = w_*^3,\qquad (12.7)$$

where h is the height of the unstable boundary layer and w_* is the convective velocity scale. We get the following from Eqs. (12.5)–(12.7):

$$k(w_*/u_*)^3 = -h/L.\qquad (12.8)$$

For more information on the above, see van Ulden and Holtslag (1985) and Holtslag and Nieuwstadt (1986).

12.2 Vertical velocity variance

We seek a parameterization for the vertical velocity variance σ_w^2 that gives continuous values for the boundary layer at all elevations ($z_0 \le z \le h$) and all stability conditions from unstable to stable ($-\infty < L < \infty$). We noted in section 12.1 that the friction velocity u_* and convective scale velocity w_* are the scale velocities for different stability regimes so we assume a general relationship of the form

$$\sigma_w^2 = f_S u_*^2 + f_B w_*^2,\qquad (12.9)$$

where f_S and f_B are shear-stress and buoyancy-force functions, respectively, to be determined. This relation is similar to Eq. (4) in Hicks (1985); see also the first part of Eq. (8) in that paper. From Eqs. (12.8) and (12.9),

$$(\sigma_w^2/u_*)^2 = f_S + f_B[h/(-kL)]^{2/3}.\qquad (12.10)$$

We propose that the second term with f_B be negligible or zero in near-neutral and stable conditions for which u_* is the appropriate scaling velocity. We also propose that the first term with f_S be negligible in very unstable conditions and at elevations for which w_* is the appropriate scaling velocity. The following could meet these criteria:

$$\left(\frac{\sigma_w}{u_*}\right)^2 = C_1\left(1 - \frac{z}{h}\right)^{3/2}$$
$$+ C_2\left[\left(\frac{z}{h}\right)\left(\frac{h}{-L}\right)\right]^{2/3}\left(1 - C_3\frac{z}{h}\right)^2,$$
$$(12.11)$$

where the constants $C_{1,2,3}$ are to be determined. The first term on the right (with C_1) is a modification of Eq. (38) in Nieuwstadt (1984). The second term (with C_2 and C_3) is based on Eq. (13) for the convective boundary layer in Lenschow et al. (1980).

Nieuwstadt (1984) found that

$$\left(\frac{\sigma_w}{u_*}\right)^2 = 1.96\left(1 - \frac{z}{h}\right)^{3/2}\qquad (12.12)$$

gave a reasonable fit to both his Cabauw tower data and the data of Caughey et al. (1979) for the stable boundary layer as shown in Nieuwstadt's Fig. 11.

For the unstable surface layer, $0 < z/h < 0.1$ with the result that the quantities $(1 - z/h)$ and $(1 - C_3z/h)$ are approximately equal to unity. We therefore approximate our Eq. (12.11) as

$$\left(\frac{\sigma_w}{u_*}\right) = C_1 + C_2\left(\frac{z}{-L}\right)^{2/3},\qquad (12.13)$$

which is identical in form to Eq. (7) in Panofsky et al. (1977):

$$\left(\frac{\sigma_w}{u_*}\right) = 1.4 + 2.9\left(\frac{z}{-L}\right)^{2/3}.\qquad (12.14)$$

Equations (12.11) and (12.13)–(12.14) are consistent with the local free convection prediction for the very unstable surface layer that $(\sigma_w/u_*)^2$ is proportional to $[z/(-L)]^{2/3}$ (see Fig. 1 in Wyngaard et al., 1971). We should note that the data for $(\sigma_w/u_*)^2$ in Panofsky et al. are significantly higher than datasets from Kansas and Sweden (see the text on p. 1959 and Fig. 16 in Högström 1990).

We noted that the term with C_2 in Eq. (12.11) is based on Eq. (13) for the convective boundary layer in Lenschow et al. (1980):

$$\left(\frac{\sigma_w}{w_*}\right)^2 = 1.8\left(\frac{z}{h}\right)^{2/3}\left(1 - 0.8\frac{z}{h}\right)^2.\qquad (12.15)$$

This equation is a fit to data from flight measurements. According to Weil (1990) p. 513, the average value of the ratio u_*/w_* for these data was 0.25. From Eq. (12.8) with $k = 0.4$, this corresponds to $h/L = -25.6$, a condition for which the effect of the shear stress is negligible except near the surface. We combine Eqs. (12.8) and (12.15) to get

$$\left(\frac{\sigma_w}{u_*}\right)^2 = 3.3\left[\left(\frac{z}{h}\right)\left(\frac{h}{-L}\right)\right]^{2/3}\left(1 - 0.8\frac{z}{h}\right)^2.$$
$$(12.16)$$

Unlike Eq. (12.11), Eqs. (12.15) and (12.16) do not account for the effect of shear stress. Weil (1990) took shear stress into account in

$$\left(\frac{\sigma_w}{w_*}\right)^2 =$$

$$\left[\left(\frac{u_*}{w_*}\right)^3 \left(1.6 - \frac{z}{h}\right)^{3/2} + 1.2\frac{z}{h}\left(1 - 0.98\frac{z}{h}\right)^{3/2}\right]^{2/3}$$

$$(12.17)$$

and fitted it to both the flight data of Lenschow et al. and large–eddy simulation data from Moeng and Wyngaard (1984, 1989). The ratio $u_*/w_* = 0.3$ in these simulations corresponds to $h/L = -14.8$.

The above are sample parameterizations and do not form a complete listing of those published. The different forms of the equations and the several values of the empirical constants serve to indicate the problems that will be encountered in developing parameterizations that will give *continuous* values of σ_w^2 over the full range of height and stability conditions. Hicks (1985) discussed at length the problems of fitting (and even constraining) data to equations like Eq. (12.9) for the entire atmospheric boundary layer.

12.3 Dissipation of turbulence kinetic energy

The parameterization for the mean rate of dissipation of turbulence kinetic energy ε is more difficult than it is for the vertical variance σ_w^2. This is in part because the dimensionless quantity $\varepsilon k h/u_*^3$ does not follow similarity laws. In addition, there are fewer sources of data, that is, field experiments and numerical simulations. We propose a pair of parameterizations:

$$\frac{\varepsilon k h}{u_*^3} = \frac{h}{z}\left[1 + C_4\left(\frac{z}{h}\right)\left(\frac{h}{L}\right)\right]\left(1 - C_5\frac{z}{h}\right)^{3/2}$$

$$(12.18)$$

for the neutral to stable case, and

$$\frac{\varepsilon k h}{u_*^3} =$$

$$"\frac{h}{z}\left(\left[1 + C_6\left(\frac{z}{h}\right)\left(\frac{h}{-L}\right)\right]\left(1 - C_5\frac{z}{h}\right)^{3/2}\right)$$

$$+ C_7\frac{h}{-L}$$

$$(12.19)$$

for unstable to neutral conditions. The first parameterization is from Brost and Wyngaard (1978), and the other is ad hoc.

Brost and Wyngaard (1978) studied the stably stratified boundary layer with a second-order turbulence model in which the rate of dissipation of the turbulence kinetic energy was assumed to be proportional to the 3/2 power of the turbulence kinetic energy divided by a length scale given by a height-dependent interpolation formula. They fitted Eq. (12.18) with the constants $C_4 = 3.7$ and $C_5 = 0.85$ to the dissipation rate calculated by their model. Lenschow et al. (1988) compared the predictions of this equation with their flight data as well those of Caughey et al. (1979) and concluded that the agreement was "generally pretty good." However, the data have considerable scatter about the values given by Eq. (12.18). Mason and Derbyshire (1990) compared the Brost and Wyngaard results for ε with theirs from large eddy simulations. They found good agreement for the stable case (for which the Brost and Wyngaard model was designed) and poor agreement for the neutral case.

We know of no counterpart to Eq. (12.18) for the unstable to neutral case. We begin our construction of Eq. (12.19) with Eq. (20) in Wyngaard and Coté (1971) for the unstable surface layer:

$$\frac{\varepsilon k z}{u_*^3} = \left(1 + 0.5\frac{z}{-L}\right)^{3/2}. \qquad (12.20)$$

The power-series expansion of the above is

$$\frac{\varepsilon k z}{u_*^3} = 1 + 0.75\frac{z}{-L} + \dots \qquad (12.21)$$

(i.e., $C_6 = 0.75$), which is a fair approximation to the curves for both the Kansas data (Wyngaard and Coté) and the Swedish data for the unstable-to-neutral surface layer plotted in Fig. 3 of Högström (1990). However, this equation is not in agreement with Högström's finding that the local rate of dissipation of turbulence kinetic energy in the neutral surface layer is about 20% greater than its local rate of production; that is, $\varepsilon k z/u_*^3 = 1.24$ instead of the traditional value of unity in neutral conditions. We acknowledge this fact, but we use $\varepsilon k z/u_*^3 = 1$ for convenience.

We ensure continuous values for $\varepsilon k z/u_*^3$ for $0 \leq z/h \leq 1$ across the unstable-to-stable transition (where $L \to \pm\infty$) if we use the z/h function with C_5 that is in Eqs. (12.18) and (12.19) with the Brost and Wyngaard value of $C_5 = 0.85$:

$$\frac{\varepsilon k z}{u_*^3} = \left(1 + 0.75\frac{z}{-L}\right)\left(1 - 0.85\frac{z}{h}\right)^{3/2}. \qquad (12.22)$$

Finally, as indicated in Fig. 5 of Luhar and Britter (1989) and Fig. 14d of Weil (1990), $\varepsilon h/w_*^3$ is, to a good degree, approximately constant with height throughout much of the convective boundary layer. Therefore, to account for the effect of buoyancy,

$$\frac{\varepsilon k z}{u_*^3} = \left(1 + 0.75\frac{z}{-L}\right)\left(1 - 0.85\frac{z}{h}\right)^{3/2} + 0.3\frac{z}{-L},$$

(12.23)

where the value of $C_7 = 0.3$ is from Weil.

12.4 Universal constant for the Lagrangian structure function

The above parameterizations for turbulence kinetic energy dissipation ε are incomplete without a numerical value for C_0, the universal constant for the Lagrangian structure function. As we noted in section 3.3, a survey of estimated values (Rodean 1991) yielded values that ranged from $C_0 = 1.6\pm0.6$ to 2.1 to 2–5 to 5–10. Pope (1994) concluded that C_0 is in the range 5.0–6.5, and Du et al. (1995) recommended $C_0 = 3.0 \pm 0.5$. Rodean obtained a semianalytical value of 5.7, which is within the range of most of these estimates.

12.5 Skewness of the vertical velocity

The following was proposed for the third moment of the vertical velocity fluctuations by Sorbjan (1990) in Eq. (17) and Weil (1990) in Eq. (36b) in the CBL ($h/L < 0$):

$$\frac{\langle w^3 \rangle}{w_*^3} = C_8 \frac{z}{h}\left(1 - \frac{z}{h}\right).$$

(12.24)

In Sorbjan, $C_8 = 0.5$, and $C_8 = 0.84$ in Weil. The two parts of Eq. (2.10) in Sawford and Guest (1987) combine to give $C_8 = 0.92$. In their paper, the right side of Eq. (12.24) is multiplied by a function of z/h that is greater than unity of $z/h < 0.3$ and less than unity if $z/h > 0.3$. The combination of Eqs. (12.8) and (12.24) with Weil's value for C_8 gives

$$\frac{\langle w^3 \rangle}{u_*^3} = \frac{0.84}{k}\left(\frac{h}{-L}\right)\frac{z}{h}\left(1 - \frac{z}{h}\right).$$

(12.25)

We note that Eq. (12.25) is not in agreement with Högström (1990) where $\langle w^3 \rangle/u_*^3 = 0.3 \pm 0.03$ for the neutral atmospheric surface layer. We have sufficient data to define non-Gaussian turbulence for *only* the unstable convective boundary layer (Luhar and Britter 1989; Weil 1990; Du et al. 1994).

12.6 Horizontal velocity variances

We suggest the following for the horizontal velocity variances σ_u^2 and σ_v^2:

$$\left(\frac{\sigma_u}{u_*}\right)^2 =$$
$$C_9\left(1 - \frac{z}{h}\right)^p + C_{10}\left(\frac{h}{-L}\right)^{2/3} = \left(\frac{\sigma_v}{u_*}\right)^2,$$

(12.26)

with $C_9 = 4.5$, $C_{10} = 0$ if $L \geq 0$, $C_{10} = 0.6$ if $L \leq 0$, and $p = 3/2$. We assume in the above that the longitudinal and lateral velocity fluctuations are equal. There are data, including those in Table 2 of Högström (1990), that indicate $\sigma_u^2 > \sigma_v^2$, but we will follow Lenschow et al. (1988) in which $\sigma_u^2 = \sigma_v^2$. We note that Eq. (12.26) has the same form as Eq. (12.11) with $C_3 = 0$. The justifications for Eqs. (12.11) and (12.26) are similar.

The term in Eq. (12.26) with C_9 is a modification of Eq. (30) for the stably stratified boundary layer in Lenschow et al. They used $p = 7/4$ but admitted that they could not distinguish between $p = 7/4$ and $p = 3/2$ in fitting their data. The use of $p = 3/2$ is consistent with Nieuwstadt's (1984) version of Eq. (12.15). Equations (12.11) and (12.26) are consistent with some, but not all, of the data shown in Fig. 5 of Nicholls and Readings (1979). Their data for the slightly unstable surface layer ($0 < z/h < 0.1$) and the near-neutral upper layer ($0.1 < z/h < 1$) shown in the top part of Fig. 12.1. Their class A and class B data are for average values of $h/L = -3.9$ and $h/L = -1.5$, respectively.

For the unstable surface layer ($0 < z/h < 0.1$), Eq. (12.26) aproximates the form of Eq. (5) (Panofsky et al., 1977):

$$\left(\frac{\sigma_u}{u_*}\right)^2 = 4 + 0.6\left(\frac{h}{-L}\right)^{2/3} = \left(\frac{\sigma_v}{u_*}\right)^2.$$

(12.27)

Panofsky et al. noted that σ_u^2 and σ_v^2 become independent of u_* and scale with w_* for large negative values of z/L. We demonstrate this by combining Eqs. (12.8) and (12.27) to give, using $k = 0.4$,

$$\left(\frac{\sigma_u}{w_*}\right)^2 = 4\left(\frac{u_*}{w_*}\right)^2 + 0.33 = \left(\frac{\sigma_v}{w_*}\right)^2.$$

(12.28)

The above is similar to the second part of Eq. (8) in Hicks (1985). For sufficiently small values of u_*/w_* (that is, sufficiently large negative values of h/L), the above is approximately equal to Eq. (5) in Caughey and Palmer (1979):

$$\left(\frac{\sigma_u}{w_*}\right)^2 = 0.35 = \left(\frac{\sigma_v}{w_*}\right)^2.$$

(12.29)

12.7 Shear stress

The realizability condition for Reynolds shear stress, τ_{uw}, is given by Eq. (1b) in Schumann (1977):

$$\tau_{uw}^2 \leq \sigma_u^2 \sigma_w^2. \qquad (12.30)$$

A parameterization for τ_{uw} will have to meet this condition. We propose

$$\frac{\tau_{uw}}{u_*^2} = \left(1 - \frac{z}{h}\right)^{3/2-q}, \qquad (12.31)$$

where

$$q(L \geq 0) = 0 \text{ and } q(L < 0) = -\frac{h/L}{2}\left(1 - \frac{h}{L}\right)^{-1}. \qquad (12.32)$$

For neutral and stable conditions, Eq. (12.31) is identical to Nieuwstadt's (1984) Eq. (36). For very unstable conditions $(h/L \ll 0)$,

$$q \approx 1/2 \text{ and } \frac{\tau_{uw}}{u_*^2} \approx 1 - \frac{z}{h} \qquad (12.33)$$

as Wyngaard et al. (1974) found in their numerical simulations of the atmospheric boundary layer.

12.8 Unfinished business

The above parameterizations in sections 12.2–12.7 are intended only to demonstrate one of many approaches that could be taken in generating input parameterizations for the LEM and RDM. The literature cited is a selected sample, not a comprehensive survey. One of the recent surveys of field observations was published by Hess (1992). We make two comments about quality data for use in parameterizations. First, there are insufficient sets of high quality data for the height and stability scaling regimes in the Holtslag–Nieuwstadt parameter space illustrated in Fig. 12.1, especially for the near-neutral upper layer. Field experiments that yield high quality data are extremely valuable, but they are very costly in terms of both time and money. We believe that high-resolution, high–Reynolds number numerical simulations will eventually provide most of the needed data.

We have not addressed the necessity of modifying the input parameterizations near boundaries so as to form the following combination (see chapter 11):

1) an LEM or a RDM,

2) boundary condition algorithms, and

3) parameterizations of inputs that meet the well-mixed criterion of Thomson (1987).

Each user will have to develop the combination of stochastic Lagrangian diffusion model, boundary conditions, and inputs that are appropriate. For some, it will be a complex model for research with detailed descriptions of the flow field and its turbulence statistics. For others, it will be a simpler model with fewer inputs for real time (or faster) simulations for use in emergency response. In any case, it will not be a trivial task.

Chapter 13

Brief Review of Four Topics

In preceding chapters, we reviewed the present state of knowledge for applying stochastic models of turbulent diffusion to the dispersion of passive, scalar tracers in the atmospheric boundary layer. Specifically, this involves the numerical integration of a stochastic differential equation to define Lagrangian trajectories of a very large population of "marked" particles in turbulent flow. In this chapter, we briefly consider four subjects of related recent and current research. We do not go into details as extensively as before, with the exception of Kolmogorov's (1962) refined hypotheses and fractional Brownian motion. We cite recent (mostly 1990 and later) literature as recommended reading.

13.1 Two-particle dispersion

Thus far, we have considered only equations that model the random trajectory of a *single* particle. The mean concentration at a point in space and an instant in time is calculated from the trajectory statistics of thousands to tens of thousands of *individual* particles. Stochastic models for Lagrangian *pairs* of particles are models of relative diffusion, and the statistics of such trajectories are closely related to concentration variance.

Thomson (1990) extends his classic 1987 paper on one-particle stochastic models to particle pairs. Borgas and Sawford (1991) examined the small-scale structure of acceleration correlations for both one-particle and two-particle statistics, with emphasis on the latter. In Borgas and Sawford (1994b), they built upon Thomson (1990) in describing the philosophy and methodology they used in developing a recommended two-particle model for use in applications. Section 2 in their paper is an excellent, concise summary of the theory of one-particle models of the stochastic Lagrangian velocity process.

A different approach to two-particle stochastic models of relative dispersion was taken by Novikov (1986, 1989) and Pedrizzetti and Novikov (1994). Pedrizzetti and Novikov considered stochastic Lagrangian modeling of the relative motion of two fluid particles in the inertial range of turbulence.

They developed a Lagrangian model for relative velocity [analogous to Eqs. (10.13a,b)] and related it to an equation for the Eulerian probability density $P_E(\mathbf{u}|t,\mathbf{r})$, the probability density at time t of the velocity difference $\mathbf{u}$ between two fixed points at distance $\mathbf{r}$. This equation for P_E introduces a hierarchy of constraints that make the model consistent with the theory of locally isotropic turbulence. They showed that their Lagrangian model, with local relaxation and Gaussian forcing, is consistent with incompressibility, classical similarity scaling, the Eulerian equation for the third-order structure function, and Eulerian data for second-order turbulence statistics.

In spite of the recent papers summarized above, the development and application of models for two-particle dispersion are not as advanced as those for single-particle models.

13.2 Intermittency and turbulent diffusion

Sections 3.2 and 3.3 summarize Kolmogorov's (1941) hypotheses for the local properties of high–Reynolds number turbulence, which are used to justify the use of the Langevin equation model of turbulent diffusion. One of the parameters in this model is the *mean* rate of turbulence energy dissipation represented by the symbol ε in all the preceding. As implied by the adjective "mean," it is more correctly represented by $\langle \varepsilon \rangle$ in the same sense that $U_E = \langle u_E \rangle$ in Eq. (8.1). However, the rate of dissipation of turbulence kinetic energy is actually a random function of space and time that fluctuates with the turbulent velocity field. In the following, ε_r is used to represent the *locally averaged* rate of dissipation over a small volume of linear dimension r. The term "intermittency" is applied to the extreme variations of ε_r (and other turbulence statistics). [See the footnote on p. 126 of Landau and Lifshitz (1959), Monin and Yaglom (1975) section 25, Lesieur (1990) section VI.8, and McComb (1990) sections 3.2.1–3.2.2.] There have been many investigations, experimental and theoretical, of the Eulerian aspects of intermittency corrections to Kolmogorov's (1941) hypotheses.

The implications of intermittency with respect to stochastic Lagrangian models of turbulent phenomena have been addressed recently by Novikov (1989, 1990), Pope and Chen (1990), Borgas (1993), Sawford (1993), and Borgas and Sawford (1994a). For our purposes, the last sentence of the abstract in Borgas and Sawford is pertinent: "The practical impact of intermittency upon dispersion is found to be small."

13.3 Kolmogorov's refined hypotheses

Even though intermittency does not have a significant effect upon turbulent diffusion, it is appropriate to review Kolmogorov's (1962) revision of his original (1941) hypotheses because the latter provide the physical justification for our stochastic Lagrangian models of turbulent diffusion (see Chapter 3). Kolmogorov based his refined hypotheses on a proposal by Obukhov (1962) that the global average of the turbulence kinetic energy dissipation rate ε be replaced by local averages ε_r, where ε_r is the average of ε over a volume of linear dimension r.

We follow Stolovitzky and Sreenivasan (1994) in restating the original (1941) hypotheses of Kolmogorov in a manner different from that given in sections 3.2–3.3. The first hypothesis is that, for the scales of motion $r \ll L$, the turbulence statistics are determined uniquely by ε and the kinematic viscosity ν. Here L is the scale at which the energy, which maintains the turbulence, is injected. The following scales of length and velocity, respectively, are determined by ε and ν:

$$\eta_K = (\nu^3/\varepsilon)^{1/4} \text{ and } u_K = (\nu\varepsilon)^{1/4}, \qquad (13.1)$$

as in Eq. (3.3). As we showed in Eq. (3.4), these scales define a Reynolds number that is equal to unity. The second (original) hypothesis is that, in the inertial subrange defined in terms of the length scales as $\eta_K \ll r \ll L$, the turbulence statistics are independent of the viscosity ν and are a function of only the mean dissipation rate ε.

We use Stolovitzky and Sreenivasan's description of Kolmogorov's (1962) refined hypotheses. We define an incremental velocity change Δu over a scale length r:

$$\Delta u(r) = u(x + r) - u(x), \qquad (13.2)$$

where $u(x)$ is the velocity component along the coordinate x and r is an incremental length along x. (Note that here we use Eulerian coordinates, in contrast to sections 3.2–3.3 where we worked in a Lagrangian system.) The first refined hypothesis may be stated in terms of the probability density function of the stochastic variable V, where

$$V = \Delta u(r)/(r\varepsilon_r)^{1/3}. \qquad (13.3)$$

The quantity $(r\varepsilon_r)^{1/3}$ is a scale velocity. Equation (13.3) applies over a range of length scales r such that $r \ll L$, and the probability density function of V is a function of the local Reynolds number,

$$\mathrm{Re_r} = r(r\varepsilon_r)^{1/3}/\nu. \qquad (13.4)$$

The second refined hypothesis is that, for $\mathrm{Re_r} \gg 1$, the probability density function of V is independent of the local Reynolds number, and is therefore universal. This means that the probability density function is not only independent of ν, as in the original second hypothesis, but is also independent of r and ε_r. (We do not consider Kolomogorov's third hypothesis of 1962.)

These refined hypotheses have been met with skepticism. Kraichnan (1974) noted that Eq. (13.3) is a mixture of quantities from two different ranges of turbulence: $\Delta u(r)$ is an inertial quantity if r is an inertial range separation distance, and ε_r is the integral of a dissipation range quantity over an inertial range scale. Frisch (1991) noted Kraichnan's observation and added objections of his own. The argument has continued. Experimental results by Praskovsky (1992), Stolovitzky et al. (1992), and Thoroddsen and Van Atta (1992) were published in support of these hypotheses, but later Thoroddsen had second thoughts.

Thoroddsen (1995) began by noting that Kolmogorov's (1962) formulation of Eq. (13.6) is a volume integral and that, in analyzing experimental data, this volume integral is replaced by an integral along a straight line r connecting the two points used to evaluate $\Delta u(r)$. Furthermore, the dissipation along r is approximated by the streamwise gradient of the streamwise velocity component, the "pseudodissipation" defined as

$$\varepsilon_r^+ = 15\nu(\partial u/\partial x)^2. \qquad (13.5)$$

Thoroddsen recalled that Thoroddsen and Van Atta (1992) had noted that this dependence is to be expected from the analytical constraint of the continuity of u along x, raising a question about the significance of the dynamics of turbulence. Thoroddsen went on to present another relation for the pseudodissipation,

$$\varepsilon_r^* = 7.5(\partial w/\partial x)^2, \qquad (13.6)$$

where w is a transverse velocity component. His analysis of experimental data showed strong correlation between Δu and ε_r^+ but little correlation between Δu and ε_r^*. These results do not disprove the refined hypotheses, but they demonstrate a weakness in the earlier reported experimental support. In a companion numerical simulation effort, "Is the Kolmogorov refined similarity relation dynamic or kinematic?" Chen et al. (1995) concluded that the correlation between Δu and ε_r^+ is mostly due to kinematics, but that there appears to be a small dynamic contribution. Mandelbrot (1982) p. 103 may have been prophetic: "Obukhov (1962) and Kolmogorov (1962) are the first analytic studies of intermittency. In immediate influence, they nearly matched the 1941 papers of the same authors, but they are seriously flawed and their long term influence promises to be small."

Let us close with the meanings of "kinematic" and "dynamic" used in the preceding paragraph. The following definitions are from the *The New Shorter Oxford English Dictionary* (1993):

Dynamic (adjective): "Of or relating to force producing motion."

Kinematic (adjective): "Of or pertaining to kinematics; relating to pure motion, considered without reference to force or mass."

In our context of fluid mechanics, the continuity equation (or equation for conservation of mass) is kinematic and the Navier–Stokes equation is dynamic.

13.4 Fractional Brownian motion

Stolovitzky and Sreenivasan (1994) stated that it is difficult to see how much of the content of Kolmogorov's refined hypotheses is specific to fluid turbulence. The intent of their paper is to demonstrate that "...a large part of the refined hypotheses can be derived from general stochastic principles unrelated to the Navier–Stokes equations."

They normalized all lengths by the Kolmogorov scale η_K and all velocities by the Kolmogorov velocity u_K. After some mathematical manipulations, they obtained the following form of Eq. (13.3):

$$V_{fBm} = \frac{\Delta Z(r)}{w_r^H} = \frac{K^{1-H}}{15^H} \frac{S_p}{Y_p^{2H}}. \qquad (13.7)$$

Here V_{fBm} is the stochastic variable V for fractional Brownian motion, $\Delta Z(r)$ is the incremental motion,

w_r is a dissipation parameter that coincides with $r\varepsilon_r$ in the case of turbulence, K is the number of Kolmogorov scale lengths η_K over which smoothness obtains, $S_p \approx \Delta Z(r)/K$, $Y_p^2 = w_r/15K$, and H is the Hurst exponent where $0 < H < 1$.

We define first the Hurst exponent and then fractional Brownian motion. The Hurst exponent (Feder 1988, chapter 8) came from Hurst's study of long-term time series from observations of natural phenomena including the statistics of water storage in lakes and reservoirs along the Nile, rainfall, temperatures, thicknesses of sedimentary layers and tree rings, etc. He developed the technique of rescaled range analysis and found that many natural time series are very well described by the following empirical relation:

$$R/S = (\tau/2)^H, \qquad (13.8)$$

where τ is the duration of the observations considered, R is the difference between the maximum and minimum observed values during the time τ, and S is the standard deviation of the observations during the same period. Equation (13.8) shows that the ratio R/S increases with duration τ. If the time series represents a statistically independent random process with finite variance, $H = 1/2$. Hurst found from the data for many natural time series that $H = 0.73$ with a standard deviation of about 0.09. This value demonstrates that there are "memory" effects in natural processes such as water discharges from a reservoir or lake in a large river system. Hurst did not name the exponent after himself; a footnote on p. 153 of Feder states that Mandelbrot used H for this exponent.

The process of fractional Brownian motion was introduced by Mandelbrot and Van Ness (1968). The following is based on their paper and Feder (1988) sections 9.4–9.5. The mean of the fractional Brownian motion process is

$$\langle B_H(t) - B_H(s) \rangle = 0, \qquad (13.9)$$

and the variance of the incremental process is

$$\langle [B_H(t) - B_H(s)]^2 \rangle = (t - s)^{2H}, \qquad (13.10)$$

where $t > s$ and $0 < H < 1$. Consider the correlation of two increments, $[B_H(s) - B_H(-t)]$ and $B_H(t) - B_H(s)$. For convenience, let $s = 0$ and $B_H(s) = 0$. Then the correlation coefficient is

$$C(t) =$$

$$\frac{\langle [B_H(s) - B_H(-t)][B_H(t) - B_H(s)] \rangle}{\langle [B_H(t) - B_H(s)]^2 \rangle} =$$

$$2^{2H-1} - 1.$$

(13.11)

If $H = 1/2$, $C(t) = 0$ and the increments of $B_H(t)$

are independent and uncorrelated. If $1/2 < H < 1$ as found by Hurst, $C(t) > 0$ and the process is *persistent*. That is, if $B_H(t)$ has tended to increase in the past, it will tend to do so in the future. Similarly, if it has tended to decrease in the past, it will tend to decrease in the future. If $0 < H < 1/2$, $C(t) < 0$ and the process is *antipersistent*. That is, if $B_H(t)$ has tended to increase in the past, it will tend to decrease in the future, and vice versa. The formal definition of fractional Brownian motion is the integral,

$$B_H(t) - B_H(0) = \frac{1}{\Gamma(H + 1/2)} \left(\int_{-\infty}^{0} [(t-s)^{H-1/2} - (-s)^{H-1/2}] dB(s) + \int_{0}^{t} (t-s)^{H-1/2} dB(s) \right), \quad (13.12)$$

where $\Gamma(H + 1/2)$ is the gamma function and $dB(s)$ is an increment of the classical Brownian motion process [that is, $dB(s) \equiv dW(t)$, the incremental Wiener process]. If $H = 1/2$, the fractional Brownian process $B_H(t)$ reduces to the Markovian Wiener process $W(t)$ that we studied in section 4.2. In contrast, if $H \neq 1/2$, the fractional Brownian process is *not* Markovian because the "past" will influence the "future."

Stolovitzky and Sreenivasan (1994) developed a formalism for computing the conditional probability distribution function for V_{fBm} as defined by Eq. (13.7). They used this procedure to show that their reformulation of Kolmogorov's refined hypotheses is applicable to classical Brownian motion and antipersistent fractional Brownian motion, but not to persistent fractional Brownian motion. They found a limited analogy between antipersistent fractional Brownian motion with $H = 1/3$ and Eulerian

turbulence data that was analyzed by Stolovitzky et al. (1992). Stolovitzky and Sreenivasan closed by commenting that Kolmogorov's refined hypotheses may become theorems of the general theory of stochastic processes.

Shao (1995) proposed a generalized Langevin equation in which the incremental Wiener process is replaced by an incremental fractional Brownian motion process. This fractional Langevin equation, with the Hurst exponent $H = 1/2$, reduces to the familiar Langevin equation. Shao presented theoretical analyses, numerical tests, and comparisons between simulations and observations that demonstrate that, with $H = 1/3$, this model reproduces the basic statistical features of homogeneous Eulerian turbulence. Shao proposed that this model be used for turbulent diffusion of nonpassive (e.g., heavy) particles.

Chapter 14

In Conclusion

We have come full circle in a study of stochastic Lagrangian models of turbulent diffusion. We began with Brownian motion as the classic stochastic process and Kolmogorov's hypotheses on the small-scale properties of turbulence. We ended with Kolmogorov's refined hypotheses and fractional Brownian motion. Along the way, we learned about the Markov and Wiener processes, integration of stochastic differential equations, Gaussian and non-Gaussian probability distribution functions, the related Langevin and Fokker–Planck equations, the well-mixed criterion, nonuniqueness of models in more than one dimension, boundary conditions, and parameterization of turbulence statistics for use as model inputs.

For turbulent diffusion in the vertical dimension, the mathematical theory appears to be quite complete (especially for Gaussian turbulence), and input parameterizations are available for *some* of the height-and-stability regimes (e.g., the surface layer). There is a scarcity of high quality turbulence statistics for use in formulating generalized inputs to the models. Specifically, data are needed for use in developing parameterizations that give *continuous* values of model inputs over the *entire* height versus stability parameter space of the atmospheric boundary layer.

There has been some progress in modeling diffusion in two-dimensional Gaussian turbulence, but only a limited amount in multidimensional, non-Gaussian turbulence. These subjects are virtually unexplored.

More work must be done on boundary conditions.

We close with some remarks on "verification." We can test our models of turbulent diffusion by comparing numerical results with 1) analytical solutions and 2) high quality data from experiments in wind tunnels and at selected sites in the field. Can we "verify" these models? Oreskes et al. (1994) answered this question in the negative because natural systems are open, not closed, and model results are not unique. Therefore, all we can hope for is partial confirmation that our models are a satisfactory approximation to incompletely understood natural phenomena.

Acknowledgments. This work was performed under the auspices of the U.S. Department of Energy by the Lawrence Livermore National Laboratory under Contract W-7405-Eng-48.

References

Arnold, L., 1974: *Stochastic Differential Equations: Theory and Applications.* John Wiley and Sons, 228 pp.

Baerentsen, J.H., and R. Berkowicz, 1984: Monte Carlo simulation of plume dispersion in the convective boundary layer. *Atmos. Environ.* **18**, 701–712.

Barnett, S., 1990: *Matrices—Methods and Applications.* Clarendon Press, 450 pp.

Borgas, M.L., 1993: The multifractal Lagrangian nature of turbulence. *Phil. Trans. Roy. Soc. London A,* **342**, 379–411.

——, and B.L. Sawford, 1991: The small-scale structure of acceleration correlations and its role in the statistical theory of turbulent dispersion. *J. Fluid Mech.,* **228**, 295–320.

——, and ——, 1994a: Stochastic equations with multifractal random increments for modeling turbulent dispersion. *Phys. Fluids,* **6**, 618–633.

——, and ——, 1994b: A family of stochastic models for two-particle dispersion in isotropic homogeneous stationary turbulence. *J.Fluid Mech.,* **279**, 69–99.

Boughton, B.A., J.M. Delaurentis, and W.E. Dunn; 1987, 1989: A stochastic model of particle dispersion in the atmosphere. *Bound.-Layer Meteor.* **40**, 147–163; **48**, 443–444.

Brost, R.A., and J.C. Wyngaard, 1978: A model study of the stably stratified planetary boundary layer. *J. Atmos. Sci.,* **35**, 1427–1440.

Calder, K.L., 1966: Concerning the similarity theory of Monin/Obukhov. *Quart. J. Roy. Meteor. Soc.* **92**, 141–146.

Caughey, S.J., and S.G. Palmer, 1979: Some aspects of turbulence structure through the depth of the convective boundary layer. *Quart. J. Roy. Meteor. Soc.,* **105**, 811–827.

——, J.C. Wyngaard, and J.C. Kaimal, 1979: Turbulence in the evolving stable boundary layer. *J. Atmos. Sci.,* **36**, 1041–1052.

Chandrasekhar, S., 1943: Stochastic problems in physics and astronomy. *Rev. Mod. Phys.,* **15**, 1–89.

Chen, S., G.D. Doolen, R.H. Kraichnan, and L.-P. Wang, 1995: Is the Kolmogorov refined similarity relation dynamic or kinematic? *Phys. Rev. Lett.,* **74**, 1755–1758.

Corrsin, S., 1962: Theories of turbulent dispersion. In *Mécanique de la Turbulence, Colloq. Intern. CNRS, Marseille,* CNRS, 470 pp.

de Baas, A.F., and I. Troen, 1989: A stochastic equation for diffusion in inhomogeneous conditions. *Physica Scripta,* **40**, 64–72.

——, H. van Dop, and F.T.M. Nieuwstadt, 1986: An application of the Langevin equation for inhomogenous conditions to dispersion in the convective boundary layer. *Quart. J. Roy. Meteor. Soc.,* **112**, 165–180.

Deardorff, J.W., and G.E. Willis, 1975: A parameterization of diffusion into the mixed layer. *J. Appl. Meteor.,* **14**, 1451–1458.

Doob, J.L., 1942: The Brownian movement and stochastic equations. *Ann. Math.,* **43**, 351–369.

——, 1953: *Stochastic Processes.* John Wiley and Sons, 654 pp.

Du, S., J.D. Wilson, and E. Yee, 1994: Probability density functions for velocity in the convective boundary layer, and implied trajectory models. *Atmos. Environ.,* **28**, 1211–1217.

——, B.L. Sawford, J.D. Wilson, and D. J. Wilson, 1995: A determination of the Kolmogorov constant (C_0) for the Lagrangian velocity structure function, using a second-order Lagrangian stochastic model for decaying homogeneous, isotropic turbulence. *Phys. Fluids,* **7**, 3083–3090.

Durbin, P.A., 1980: A random flight model of inhomogeneous turbulent diffusion. *Phys. Fluids,* **23**, 2151–2153.

——, 1983: Stochastic differential equations and turbulent dispersion. *NASA Reference Publication 1103,* 69 pp.

——, 1984: Comments on papers by Wilson *et al.* (1981) and Legg and Raupach (1982). *Bound.-Layer Meteor.,* **29**, 409–411.

——, and J.C.R. Hunt, 1980: Dispersion from elevated sources in turbulent boundary layers. *J. de Méc.* **19**, 679–695.

Feder, J., 1988: *Fractals.* Plenum Press, 283 pp.

Feller, W., 1968: *An Introduction to Probability Theory and its Applications*—Vol. 1, 3d ed., John Wiley and Sons, 509 pp.

——, 1971: *An Introduction to Probability Theory and its Applications*—Vol. 2, 2d ed., John Wiley and Sons, 669 pp.

Flesch, T.K., and J.D. Wilson, 1992: A two-dimensional trajectory-simulation model for non-Gaussian, inhomogeneous turbulence within plant canopies. *Bound.-Layer Meteor.*, **61**, 349–374.

Frisch, U., 1991: From global scaling, *à la* Kolmogorov, to local multifractal scaling in fully developed turbulence. *Proc. Roy. Soc. London A*, **434**, 89–99.

Gardiner, C.W., 1983; 2nd ed. 1990: *Handbook of Stochastic Methods.* Springer-Verlag, 442 pp.

Gifford, F.A., 1982: Horizontal diffusion in the atmosphere: A Lagrangian-dynamical theory. *Atmos. Environ.*, **16**, 505–512.

Gihman, I.I., and A.V. Skorohod, 1972: *Stochastic Differential Equations*, Springer-Verlag, 354 pp.

Gradshteyn, I.S., and I.M. Ryzhik, 1994: *Tables of Integrals, Series, and Products*, 5th ed. A. Jeffrey, Ed., Academic Press, 1204 pp.

Gray, A.H., Jr., and T.K. Caughey, 1965: A controversy in problems involving random parametric excitation. *J. Math. Phys.*, **44**, 288–296.

Hänggi, P., and P. Talkner, 1980: A remark on truncation schemes of cumulant hierarchies. *J. Stat. Phys.*, **22**, 65–67.

Hanna, S.R., 1979: Some statistics of Lagrangian and Eulerian wind fluctuations. *J. Appl. Meteor.*, **18**, 518–525.

——, 1981: Lagrangian and Eulerian time-scale relations in the daytime boundary layer. *J. Appl. Meteor.* **20**, 242–249.

Haworth, D.C., and S.B. Pope, 1986: A generalized Langevin model for turbulent flows. *Phys. Fluids*, **29**, 387–405.

Hess, G.D., 1992: Observations and scaling of the atmospheric boundary layer. *Aust. Meteor. Mag.*, **41**, 79–99.

Hicks, B.B., 1985: Behavior of turbulence statistics in the convective boundary layer. *J. Climate Appl. Meteor.*, **24**, 607–614.

Högström, U., 1990: Analysis of turbulence structure in the surface layer with a modified similarity formulation for near neutral conditions. *J. Atmos. Sci.*, **47**, 1949–1972.

Holtslag, A.A.M., and F.T.M. Nieuwstadt, 1986: Scaling the atmospheric boundary layer. *Bound.-Layer Meteor.*, **36**, 201–209.

——, and C.-H. Moeng, 1991: Eddy diffusivity and counter gradient transport in the convective atmospheric boundary layer. *J. Atmos. Sci.*, **48**, 1690–1698.

Ito, K., 1951: On stochastic differential equations. *Mem. Amer. Mathem. Soc.*, **4**, 1–51.

Jaeger, L.G., 1966: *Cartesian Tensors in Engineering Science.* Pergamon Press, 116 pp.

Kolmogorov, A.N., 1931: Uber die analytischen Methoden in der Wahrscheinlichkeitsrechnung (On analytical methods in probability theory). *Math. Ann.*, **104**, 415–458.

——, 1941: Local structure of turbulence in an incompressible viscous fluid at very high Reynolds numbers. *Dokl. Akad. Nauk.*, **30**, 301–305. Reprinted in *Sov. Phys. Usp.*, **10**, 734–736, 1968, and *Proc. Roy. Soc. London A*, **434**, 9–13, 1991.

——, 1962: A refinement of previous hypotheses concerning the local structure of turbulence in a viscous fluid at high Reynolds number. *J. Fluid Mech.*, **13**, 82–85.

Kraichnan, R.H., 1974: On Kolmogorov's inertial-range theories. *J. Fluid Mech.*, **62**, 305–330.

Krasnoff, E., and R.L. Peskin, 1971: The Langevin model for turbulent diffusion. *Geophys. Fluid Dyn.*, **2**, 123–146.

Landau, L.D., and E.M. Lifshitz, 1959: *Fluid Mechanics.* Addison-Wesley, 536 pp.

Legg., B.J., and M.R. Raupach, 1982: Markov-chain simulation of particle diffusion in inhomogeneous flows: The mean drift velocity induced by a gradient in Eulerian velocity variance. *Bound.-Layer Meteor.*, **24**, 3–13.

Lenschow, D.H., J.C. Wyngaard, and W.T. Pennel, 1980: Mean-field and second-moment budgets in a baroclinic, convective boundary layer. *J. Atmos. Sci.*, **37**, 1313–1326.

——, X.S. Li, C.J. Zhu, and B.B. Stankov, 1988: The stably stratified boundary layer over the Great Plains. *Bound.-Layer Meteor.*, **42**, 95–121.

Lesieur, M., 1990: *Turbulence in Fluids*, 2d rev. ed., Kluwer Academic Publishers, 412 pp.

Ley, A.J., 1982: A random walk simulation of two-dimensional turbulent diffusion in the neutral surface layer. *Atmos. Environ.*, **16**, 2799–2808.

——, and D.J. Thomson, 1983: A random walk model of diffusion in the diabatic surface layer. *Quart. J. R. Met. Soc.* **109**, 867–880.

Lin, C.C., and W.H. Reid, 1963: Turbulent flow, theoretical aspects. *Hand. Physik*, **VIII/2**, 438–523.

Luhar, A.K., and R.E. Britter, 1989: A random walk model for diffusion in inhomogeneous turbulence in a convective boundary layer. *Atmos. Environ.* **23**, 1911–1924.

——, and B.L. Sawford, 1995: Lagrangian stochastic modeling of the coastal fumigation phenomenon. *J. Appl. Meteor.*, **34**, 2259–2277.

——, M.F. Hibberd, and P.J. Hurley, 1996: Comparison of closure schemes used to specify the velocity PDF in Lagrangian stochastic dispersion models for convective conditions. *Atmos. Environ.* **30**, 1407–1418.

Lukacs, E., 1958: Some extensions of a theorum of Marcienkiewicz. *Pacific J. Math.* **8**, 487–501.

———, 1960: *Characteristic Functions.* Hafner Publishing Company, 216 pp.

Lumley, J.L., 1962: The mathematical nature of the problem of relating Lagrangian and Eulerian statistical functions in turbulence. In *Mecanique de la Turbulence, Colloq. Intern. CNRS, Marseille*, CNRS, 470 pp.

Maeder, R., 1991: *Programming in Mathematica*, 2d ed., Addison-Wesley, 279 pp.

Malliaris, A.G., 1983: Ito's calculus in financial decision making. *SIAM Rev.*, **25**, 481–496.

Mandelbrot, B.B., 1982: *The Fractal Geometry of Nature*, W.H. Freeman and Company, 460 pp.

———, and J.W. Van Ness, 1968: Fractional Brownian motions, fractional noises and applications. *SIAM Rev.*, **120**, 422–437.

Marcienkiewicz, J., 1939: Sur une propriètè de la loi de Gauss. *Math. Z.*, **44**, 612–618.

Mason, P.J., and S.H. Derbyshire, 1990: Large-eddy simulation of the stably-stratified atmospheric boundary layer. *Bound.-Layer Meteor.*, **53**, 117–162.

McComb, W.D., 1990: *The Physics of Fluid Turbulence.* Clarendon Press, 572 pp.

Moeng, C.-H., and J.C. Wyngaard, 1984: Statistics of conservative scalars in the convective boundary layer. *J. Atmos. Sci.*, **41**, 3161–3169.

———, and ———, 1989: Evaluation of turbulent transport and dissipation closures in second-order modeling. *J. Atmos. Sci.*, **46**, 2311–2330.

Monin, A.S., 1959: On the boundary condition on the earth surface for diffusing pollution. *Adv. Geophys.*, **6**, 435–436.

———, and A.M. Yaglom, 1971: *Statistical Fluid Mechanics: Mechanics of Turbulence*—Vol. 1. MIT Press, 769 pp.

———, and ———, 1975: *Statistical Fluid Mechanics: Mechanics of Turbulence*—Vol. 2. MIT Press, 874 pp.

Mortensen, R.E., 1969: Mathematical problems of modeling stochastic nonlinear dynamic systems. *J. Stat. Phys.*, **1**, 271–296.

Nicholls, S., and C.J. Readings, 1979: Aircraft observations of the structure of the lower boundary layer over the seas. *Quart. J. Roy. Meteor. Soc.*, **105**, 785–802.

Nieuwstadt, F.T.M., 1984: The turbulent structure of the stable, nocturnal boundary layer. *J. Atmos. Sci.*, **41**, 2202–2216.

Novikov, E.A., 1963: Random force method in turbulence theory. *Sov. Phys. JETP*, **17**, 1449–1454.

Novikov, E.A., 1986: The Lagrangian-Eulerian probability relations and the random force method for nonhomogeneous turbulence. *Phys. Fluids*, **29**, 3907–3909.

———, 1989: Two-particle description of turbulence, Markov property, and intermittency. *Phys. Fluids A*, **1**, 326–330.

———, 1990: The effects of intermittency on statistical characteristics of turbulence and scale similarity of breakdown coefficients. *Phys. Fluids A*, **2**, 814–820.

Obukhov, A.M., 1959: Description of turbulence in terms of Lagrangian variables. *Adv. Geophys.*, **6**, 113–116.

———, 1962: Some specific features of atmospheric turbulence. *J. Fluid Mech.*, **13**, 77–81.

Øksendal, B., 1992: *Stochastic Differential Equations*, 3d ed. Springer-Verlag, 224 pp.

Oreskes, N., K. Shrader-Frechette, and K. Belitz, 1994: Verification, validation, and confirmation of numerical models in the earth sciences. *Science*, **263**, 641–646.

Panofsky, H.A., H. Tennekes, D.H. Lenschow, and J.C. Wyngaard, 1977: The characteristics of turbulent velocity components in the surface layer under convective conditions. *Bound.-Layer Meteor.*, **11**, 355–361.

Papoulis, A., 1962: *The Fourier Integral and its Applications.* McGraw-Hill, 318 pp.

Pawula, R.F., 1967a: Generalizations and extensions of the Fokker-Planck-Kolmogorov equations. *IEEE Trans. Info. Theory*, **13**, 33–41.

———, 1967b: Approximation to the linear Boltzmann equation by the Fokker-Planck equation. *Phys. Rev.* **162**, 186–188.

Pedrizzetti, G., and E.A. Novikov, 1994: On Markov modelling of turbulence. *J. Fluid Mech.*, **280**, 69–93.

Pope, S.B., 1983: A Lagrangian two-time probability density function equation for inhomogeneous turbulent flows. *Phys. Fluids*, **26**, 3448–3450.

———, 1987: Consistency conditions for random-walk models of turbulent diffusion. *Phys. Fluids*, **30**, 2374–2379.

———, 1994: Lagrangian PDF methods for turbulent flows. *Ann. Rev. Fluid Mech.*, **26**, 23–63.

———, and Y.L. Chen, 1990: The velocity-dissipation probability density function model for turbulent flows. *Phys. Fluids A*, **2**, 1437–1449.

Praskovsky, A.A., 1992: Experimental verification of Kolmogorov's refined similarity hypothesis. *Phys. Fluids A*, **4**, 2589–2591.

Risken, H., 1989: *The Fokker-Planck Equation*, 2d ed. Springer-Verlag, 472 pp.

Robinson, D.W., 1965: A theorem concerning the positive metric. *Commun. Math. Phys.*, **1**, 89–94.

Rodean, H.C., 1991: The universal constant for the Lagrangian structure function. *Phys. Fluids A*, **3**, 1479–1480.

——, 1994: Notes on the Langevin model for turbulent diffusion of "marked" particles. 122 pp. *Lawrence Livermore National Laboratory Report UCRL-ID-115869.* [Available from Atmospheric Sciences Library, L-103, Lawrence Livermore National Laboratory, P.O. Box 808, Livermore, California 94550.]

Sawford, B.L., 1984: The basis for, and some limitations of the Langevin equation in atmospheric relative diffusion modelling. *Atmos. Environ.*, **18**, 2405–2411.

——, 1985: Lagrangian statistical simulation of concentration mean and fluctuation fields. *J. Climate Appl. Meteor.*, **24**, 1152–1166.

——, 1986: Generalized random forcing in random-walk turbulent diffusion models. *Phys. Fluids*, **29**, 3582–3585.

——, 1991: Reynolds number effects in Lagrangian stochastic models of turbulent dispersion. *Phys. Fluids A*, **3**, 1577–1586.

——, 1993: Recent developments in the Lagrangian stochastic theory of turbulent dispersion. *Bound.-Layer Meteor.*, **62**, 197–215.

——, and F.M. Guest, 1987: Lagrangian stochastic analysis of flux-gradient relationships in the convective boundary layer. *J. Atmos. Sci.*, **44**, 1152–1169.

——, and ——, 1988: Uniqueness and universality of Lagrangian stochastic models of turbulent dispersion. *Proc. of the Eighth Symp. on Turbulence and Diffusion. San Diego, California.* Amer. Meteor. Soc., 96–99.

——, and M.S. Borgas, 1994: On the continuity of stochastic models for the Lagrangian velocity in turbulence. *Physica D*, **76**, 297–311.

Schumann, U., 1977: Realizability of Reynolds-stress turbulence models. *Phys. Fluids*, **20**, 721–725.

Searle, S.R., 1982: *Matrix Algegra Useful for Statistics.* John Wiley and Sons, 438 pp.

Shao, Y., 1995: The fractional Ornstein-Uhlenbeck process as a representation of homogeneous Eulerican velocity turbulence. *Physica D*, **83**, 461–477.

Sorbjan, Z., 1990: Similarity scales and universal profiles of statistical moments in the convective boundary layer. *J. Appl. Meteor.* **29**, 762–775.

Spanier, J., and K.B. Oldham, 1987: *An Atlas of Functions.* Hemisphere Publishing Corporation, 700 pp.

Stolovitzky, G., and K.R. Sreenivasan, 1994: Kolmogorov's refined similarity hypotheses for turbulence and general stochastic processes. *Rev. Mod. Phys.*, **66**, 229–240.

——, P. Kailasnath, and K.R. Sreenivasan, 1992: Kolmogorov's refined similarity hypotheses. *Phys. Rev. Lett.*, **69**, 1178–1181.

Stratonovich, R.L., 1966: A new representation for stochastic integrals and equations. *J. SIAM Control* **4**, 362–371.

——, 1968: *Conditional Markov Processes and their Application to the Theory of Optimal Control.* American Elsevier Publishing Company, Inc., 350 pp.

Taylor, G.I., 1921: Diffusion by continuous moments. *Proc. London Math. Soc. Ser. 2*, **20**, 196–211.

Thomson, D.J., 1984: Random walk modelling of diffusion in inhomogeneous turbulence. *Quart. J. Roy. Meteor. Soc.*, **110**, 1107–1120.

——, 1987: Criteria for the selection of stochastic models of particle trajectories in turbulent flows. *J. Fluid Mech.*, **180**, 529–556.

——, 1990: A stochastic model for the motion of particle pairs in isotropic high-Reynolds-number turbulence, and its application to the problem of concentration variance. *J. Fluid Mech.*, **210**, 113–153.

——, and M.R. Montgomery, 1994: Reflection boundary conditions for random walk models of dispersion in non-Gaussian turbulence. *Atmos. Environ.*, **28**, 1981–1987.

Thoroddsen, S.T., 1995: Reevaluation of the experimental support for the Kolmogorov refined similarity hypothesis. *Phys. Fluids*, **7**, 691–693.

——, and C.W. Van Atta, 1992: Experimental evidence supporting Kolmogorov's refined hypothesis. *Phys. Fluids A*, **4**, 2592–2594.

Uhlenbeck, G.E., and L.S. Ornstein, 1930: On the theory of the Brownian motion. *Phys. Rev.*, **36**, 823–841.

van Dop, H., F.T.M. Nieuwstadt, and J.C.R. Hunt, 1985: Random walk models for particle displacements in inhomogeneous unsteady turbulent flows. *Phys. Fluids*, **28**, 1639–1653.

van Kampen, N.G., 1981; 2d ed., 1992: *Stochastic Processes in Physics and Chemistry.* North-Holland Publishing Company, 419 and 465 pp.

van Ulden, A.P., and A.A.M. Holtslag, 1985: Estimation of atmospheric boundary layer parameters for diffusion applications. *J. Climate Appl. Meteor.*, **24**, 1196–1207.

Wang, M.C., and G.E. Uhlenbeck, 1945: On the theory of the Brownian motion II. *Rev. Mod. Phys.*, **17**, 323–342.

Weil, J.C., 1990: A diagnosis of the asymmetry in top-down and bottom-up diffusion using a Lagrangian stochastic model. *J. Atmos. Sci.*, **47**, 501–515.

Wiener, N., 1923: Differential space. *J. Math. Phys.*, **2**, 131–174.

Wilson, J.D., and T.K. Flesch, 1992: The well-mixed constraint applied to random flight models with reflecting boundaries. *Proceedings of the 10th Symp. on Turbulence and Diffusion. Portland, Oregon..* Amer. Meteor. Soc., 222-225.

——, and ——, 1993: Flow boundaries in random-flight models: enforcing the well-mixed condition. *J. Appl. Meteor.*, **32**, 1695–1707.

——, G.W. Thurtell, and G.E. Kidd, 1981: Numerical simulation of particle trajectories in inhomogeneous turbulence Part II: Systems with variable turbulence velocity scale. *Bound.-Layer Meteor.*, **21**, 423–441.

——, B.J. Legg, and D.J. Thomson, 1983: Calculation of particle trajectories in the presence of a gradient in turbulent-velocity variance. *Bound.-Layer Meteor.*, **27**, 163–169.

——, F.J. Ferrandino, and G.W. Thurtell, 1989: A relationship between deposition velocity and trajectory reflection probability for use in stochastic Lagrangian dispersion models. *Agric. Forest Meteor.*, **49**, 139–154.

Wolfram, S., 1991: *Mathematica—A System for Doing Mathematics by Computer*, 2d ed., Addison-Wesley, 961 pp.

Wyngaard, J.C., 1987: A physical mechanism for the asymmetry in top-down and bottom-up diffusion. *J. Atmos. Sci.* **44**, 1083–1087.

——, and O.R. Coté, 1971: The budgets of turbulent kinetic energy and temperature variance in the atmospheric surface layer. *J. Atmos. Sci.* **28**, 190–201.

——, and J.C. Weil, 1991: Transport asymmetry in skewed turbulence. *Phys. Fluids* **3**, 155–162

——, O.R. Coté, and Y. Izumi, 1971: Local free convection, similarity, and the budgets of shear stress and heat flux. *J. Atmos. Sci.* **28**, 1171–1182.

——, ——, and K.S. Rao, 1974: Modelling the atmospheric boundary layer, *Adv. Geophysics* **18A**, 193–211.

Annotated Bibliography

Arnold, L., 1974: *Stochastic Differential Equations: Theory and Applications.* John Wiley and Sons. This is an advanced text written in a highly mathematical style. It presents the Langevin equation for Brownian motion as the archtype of stochastic differential equations. The Introduction is an excellent description of the need for a new mathematical theory in order that stochastic differential equations be solved in a rigorous manner. The transformation of the Langevin equation into its Markov or diffusion equation limit is presented on pp. 134–136. The emphasis is on the Ito integral, but the Stratonovich integral is treated in chapter 10.

Chandrasekhar, S., 1943: Stochastic problems in physics and astronomy. *Rev. Mod. Phys.*, **15**, 1–89. This is a massive work with much mathematical detail. Chapters I–II and the corresponding part of the "Bibliographical Notes" are recommended for study.

de Baas, A.F., and I. Troen, 1989: A stochastic equation for diffusion in inhomogeneous conditions. *Physica Scripta*, **40**, 64–72. This paper is included in this list because Part 5 presents in detail the procedure for transforming the Langevin equation into its diffusion equation limit. It is demonstrated that two forms of the Langevin equation with different "drift corrections," one that satisfies the "well-mixed" condition and another that does not, reduce to the same random displacement equation.

Doob, J.L., 1953: *Stochastic Processes.* John Wiley and Sons. This is a classic text that is still in print. The relation between continuous Markov processes and diffusion equations is the subject of chapter IV, section 3. This section includes several of Ito's contribution to the rigorous treatment of stochastic differential equations.

Gardiner, C.W., 1983: *Handbook of Stochastic Methods.* Springer-Verlag. This is an excellent, easy to understand text that gives balanced treatments of both the stochastic differential equation and the Fokker–Planck equation.

_____ 1990: *Handbook of Stochastic Methods*, 2d ed. Springer-Verlag. The principal change from the 1983 edition is a new discussion of the Ito and Stratonovich procedures for integrating stochastic differential equations.

Kolmogorov, A.N., 1941: Local structure of turbulence in an incompressible viscous fluid at very high Reynolds numbers. *Dokl. Akad. Nauk.*, **30**, 301–305. Reprinted in *Sov. Phys. Usp.*, **10**, 734–736, 1968, and *Proc. Roy. Soc. London A*, **434**, 9–13, 1991. This is perhaps one of the most significant and most cited papers in the turbulence literature as indicated by the above two (of several) reprintings. See section 16 in Lin and Reid (1963) for one of the better explanations of Kolmogorov's theory that is called the theory of local similarity, the theory of small eddies, the theory of universal equilibrium, and the theory of locally isotropic turbulence in various references.

Legg., B.J., and M.R. Raupach, 1982: Markov-chain simulation of particle diffusion in inhomogeneous flows: The mean drift velocity induced by a gradient in Eulerian velocity variance. *Bound.-Layer Meteor.*, **24**, 3–13. This is a good introduction to the Langevin equation as a stochastic differential equation. It was discovered later that their proposed "drift correction" does not satisfy the "well-mixed" condition.

Lin, C.C., and W.H. Reid, 1963: Turbulent flow, theoretical aspects. *Hand. Physik*, **VIII/2**, 438–523. This is a highly recommended reading as both a summary of knowledge of turbulence theory at that time and, in chapter VII, sections 43–45, a pioneering application of the Langevin equation to turbulent diffusion.

Luhar, A.K., and R.E. Britter, 1989: A random walk model for diffusion in inhomogeneous turbulence in a convective boundary layer. *Atmos. Environ.*, **23**, 1911–1924. The first publication of a Langevin equation with an *approximation* to non-Gaussian turbulence that satisfies the "well-mixed" condition.

Monin, A.S., and A.M. Yaglom, 1971: *Statistical Fluid Mechanics: Mechanics of Turbulence*—Vol. 1, MIT Press. An encyclopedic treatise beginning with fundamentals and concluding with chapter 5 on turbulent dispersion.

_____, and _____, 1975: *Statistical Fluid Mechanics: Mechanics of Turbulence*—Vol. 2, MIT Press. Continuation of Vol. 1 with emphasis on more advanced topics. Turbulent diffusion is discussed in chapter 8, section 24.

Øksendal, B., 1992: *Stochastic Differential Equations*, 3d ed. Springer-Verlag. This up-to-date textbook is written in the mathematical style of Arnold (1974) and Doob (1953), but is more of an introduction to the student. It includes many examples of stochastic differential equations applied to different fields. The Ito and Stratonovich integrals are discussed in some detail, with emphasis on the former.

Pawula, R.F., 1967a: Generalizations and extensions of the Fokker–Planck–Kolmogorov equations. *IEEE Trans. Info. Theory*, **13**, 33–41. This is better introductory material than the more-often cited Pawula, R.F., 1967b: Approximation to the linear Boltzmann equation by the Fokker–Planck equation. *Phys. Rev.*, **162**, 186–188. 1967a includes excellent definitions of the forward Fokker–Planck and backward Kolmogorov equations. Both 1967a and 1967b are cited principally for the proof that if a partial differential equation of the Fokker–Planck type is of finite order, it must be of order two or less. A third-order Kramers–Moyal expansion is truncated because an infinite number of nonzero higher-order terms is neglected. This is directly related to the problems of realizable nonnegative probability distributions in partial differential equations and Gaussian versus non-Gaussian stochastic terms in stochastic differential equations.

Sawford, B.L., 1985: Lagrangian statistical simulation of concentration mean and fluctuation fields. *J. Climate Appl. Meteor.*, **24**, 1152–1166. This is an excellent introductory review of theory and applications of stochastic Lagrangian models of turbulent diffusion.

——, 1986: Generalized random forcing in random-walk turbulent diffusion models. *Phys. Fluids*, **29**, 3582–3585. This and van Dop et al. (1985) are significant precursors to Thomson (1987) in providing a rigorous mathematical foundation for use of the Langevin equation as a model of turbulent diffusion.

——, and F.M. Guest, 1988: Uniqueness and universality of Lagrangian stochastic models of turbulent dispersion. *Proc. of the Eighth Symp. on Turbulence and Diffusion. San Diego, California.* Amer. Meteor. Soc., 96–99. A demonstration of Thomson's (1987) uniqueness problem with two stochastic differential equations for random velocity increments in two dimensions that meet the "well-mixed" condition but model different diffusion processes.

Searle, S.R., 1982: *Matrix Algebra Useful for Statistics.* John Wiley and Sons. This is a useful reference for definitions of determinants, inverses of a matrix (or tensor), quadratic forms, etc. associated with Thomson's (1987) "simplest" and other solutions of the Fokker–Planck equation for multidimensional diffusion in nonstationary inhomogeneous turbulence.

Thomson, D.J., 1984: Random walk modelling of diffusion in inhomogeneous turbulence. *Quart. J. Roy. Meteor. Soc.* **110**, 1107–1120. Rigorous derivations of two "drift corrections" for the Langevin are presented. Characteristic functions are used to obtain outputs (velocity fluctuations) that are mathematically consistent with inputs (stochastic forcing functions). The "drift correction" that meets the "well-mixed" condition requires Gaussian forcing. The drift correction that does not meet this condition requires a *nonrealizable* non-Gaussian forcing function with a nonzero third moment and zero values for all higher moments.

——, 1987: Criteria for the selection of stochastic models of particle trajectories in turbulent flows. *J. Fluid Mech.*, **180**, 529–556. This is a *classic* paper. Several criteria for random velocity-increment models of turbulent diffusion are examined. It is shown that the "well-mixed" condition satisfies the other proposed criteria and that it is equivalent to most of them. It is also shown that the forcing must be Gaussian. The Fokker–Planck equation is used [instead of characteristic functions as in Thomson (1984)] to obtain the correct form of the Langevin equation as a model for diffusion in nonstationary inhomogeneous turbulence. It is stated, but not demonstrated, that there is no *unique* solution for more than one spatial dimension. The simplest solution for two or three dimensions is presented but not derived.

Uhlenbeck, G.E., and L.S. Ornstein, 1930: On the theory of the Brownian motion. *Phys. Rev.*, **36**, 823–841. This is a *classic* paper as demonstrated by the term "Ornstein–Uhlenbeck process" used in textbooks (e.g., Arnold 1974; Gardiner 1983, 1990; van Kampen 1981, 1992).

van Dop, H., F.T.M. Nieuwstadt, and J.C.R. Hunt, 1985: Random walk models for particle displacements in inhomogeneous unsteady turbulent flows. *Phys. Fluids*, **28**, 1639–1653. Mathematical aspects of the Langevin and Fokker–Planck equations as models of turbulent diffusion are discussed in detail. This paper and Sawford (1986) are important precursors to Thomson (1987).

van Kampen, N.G., 1981: *Stochastic Processes in Physics and Chemistry.* North-Holland Publishing Company. This is a comprehensive and readable text with an excellent balance between mathematical theory and applications, with emphasis on the Fokker–Planck equation. The Langevin equation is treated only in the last part of chapter VIII.

——, 1992: *Stochastic processes in physics and chemistry,* 2d ed. North-Holland Publishing Company. This thorough revision has separate chapters, VIII and IX, for the Fokker–Planck and Langevin equations, respectively.

Wang, M.C., and G.E. Uhlenbeck, 1945: On the theory of the Brownian motion II. *Rev. Mod. Phys.*, **17**, 323–342. This is of more than historical interest because it presents the Fokker–Planck and Langevin equations, the relation between them, and their applications as models of Brownian motion.

Wilson, J.D., and T.K. Flesch, 1993: Flow boundaries in random-flight models: enforcing the well-mixed condition. *J. Appl. Meteor.*, **32**, 1695–1707. This is an important paper because it is the first to concentrate on boundary conditions for Langevin equation models that meet the "well-mixed" condition.

——, B.J. Legg, and D.J. Thomson, 1983: Calculation of particle trajectories in the presence of a gradient in turbulent-velocity variance. *Bound.-Layer Meteor.*, **27**, 163–169. Several "drift corrections" for one-dimensional diffusion in inhomogeneous turbulence are compared in numerical simulations. The paper closes with this comment about one correction: "Though we can offer no proof that [the best-performing] is 'correct', it seems worthy of investigation." This drift correction was rigorously derived by Thomson (1984). It was later shown by Sawford (1986) and Thomson (1987) that the selected "drift correction" meets the "well-mixed" condition.

Wyngaard, J.C., and J.C. Weil, 1991: Transport asymmetry in skewed turbulence. *Phys. Fluids*, **3**, 155–162. Numerical simulations are used to investigate the asymmetry of "top-down" and "bottom-up" transport in the convective boundary layer. It is proposed that the asymmetry is a consequence of the interaction between the skewness of the transporting turbulence and the gradient of the flux of the transported scalar.

Author Index

Arnold, L.   4–5, 10, 13–15, 27–29, 71, 77

Baerentsen, J.H.   45–49, 51, 71

Barnett, S.   36, 71

Belitz, K.   69, 73

Berkowicz, R.   45–49, 51, 71

Borgas, M.L.   8, 16, 24, 32, 40–41, 65–66, 71, 74

Boughton, B.A.   5, 28–29, 51, 54, 71

Britter, R.E.   5, 47–49, 51–52, 54, 62–63, 72, 77

Brost, R.A.   62, 71

Calder, K.L.   43, 71

Caughey, S.J.   61–62, 64, 71

Caughey, T.K.   26–27, 72

Chandrasekhar, S.   3–4, 7, 71, 77

Chen, S.   68, 71

Chen, Y.L.   66, 73

Corrsin, S.   9, 71

Coté, O.R.   62, 64, 75

Deardorff, J.W.   45, 71

de Baas, A.F.   5, 29, 45–47, 51, 71, 77

Delaurentis, J.M.   5, 28–29, 51, 54, 71

Derbyshire, S.H.   62, 73

Doob, J.L.   3, 13, 15–16, 18, 21, 23, 25–27, 35–36, 71, 77

Doolen, G.D.   67, 71

Du, S.   8, 45–46, 49, 63, 71

Dunn, W.E.   5, 28–29, 51, 54, 71

Durbin, P.A.   5, 16, 28, 31, 54, 71

Feder, J.   67, 71

Feller, W.   21, 27, 71

Ferrandino, F.J.   52, 74

Flesch, T.K.   5, 47, 52–57, 71, 75, 79

Frisch, U.   66, 72

Gardiner, C.W.   3–4, 13, 15, 18, 20, 23, 26–27, 72, 77

Gifford, F.A.   4, 72

Gihman, I.I.   54, 72

Gradshteyn, I.S.   57, 72

Gray, A.H., Jr.   26–27, 72

Guest, F.M.   40–41, 45–47, 51–52, 63, 74, 78

Hänggi, P.   18, 72

Hanna, S.R.   4, 8, 72

Haworth, D.C.   31, 72

Hess, G.D.   64, 72

Hibberd, M.F.   45–46, 48–49, 72

Hicks, B.B.   61–63, 72

Högström, U.   61–63, 72

Holtslag, A.A.M.   45, 59–61, 72, 74

Hunt, J.C.R.   5, 9–10, 13, 18–19, 28, 51, 71, 74, 78

Hurley, P.J.   45–46, 48–49, 72

Ito, K.   25–26, 72

Izumi, Y.   61, 75

Jaeger, L.G.   39, 72

Kailasnath, P.   66, 68, 74

Kaimal, J.C.   61–62, 71

Kidd, G.E.   4, 7, 75

Kolmogorov, A.N.   1, 4, 8, 22–23, 25, 65–67, 72, 77

Kraichnan, R.H.   66, 71–72

Krasnoff, E.   4, 72

Landau, L.D.   65, 72

Legg, B.J.   4–5, 9–10, 18, 25, 28–29, 33, 44, 47, 72, 75, 77

Lenschow, D.H.   61–63, 72–73

Lesieur, M.   65, 72

Ley, A.J.   51, 72

Li, X.S.   62–63, 72

Lifshitz, E.M.   65, 72

Lin, C.C.   4–5, 9–10, 13, 72, 77

Luhar, A.K.   5, 45–49, 51–52, 54, 62–63, 72, 77

Lukacs, E.   18, 72

Lumley, J.L.   9–10, 73

Maeder, R.   14, 73

Malliaris, A.G.   27, 73

Mandelbrot, B.B.   67, 73

Marcienkiewicz, J.   18, 23, 73

Mason, P.J.   62, 73

McComb, W.D.   9, 65, 73

Moeng, C.-H.   45, 62, 73

Monin, A.S.   8–9, 54, 65, 73, 77

Montgomery, M.R.   54, 74

Mortensen, R.E.   27, 73

Nicholls, S.   63, 73

Nieuwstadt, F.T.M.   4–5, 9–10, 13, 18–19, 28, 45–47, 51, 59–61, 63, 71–74, 78

Novikov, E.A.   4, 7, 65, 73

Obukhov, A.M.   4, 13, 19, 66–67, 73

Øksendal, B.   13, 26–27, 73, 77–78

Oldham, K.B.   57, 74

Oreskes, N.   69, 73

Ornstein, L.S.   3–4, 10, 16, 74, 78

Palmer, S.G.   64, 71

Panofsky, H.A.   61, 63, 73

Papoulis, A.   14–15, 73

Pawula, R.F.   18, 23, 73, 78

Pedrizzetti, G.   65, 73

Pennel, W.T.   61–62, 72

Peskin, R.L. 4, 72

Pope, S.B.   .   .   .   8, 10, 16, 25, 31, 63, 66, 72–73

Praskovsky, A.A. 66, 73

Rao, K.S. 64, 75

Raupach, M.R.   .   .   .   4, 9–10, 18, 28–29, 47, 72, 77

Readings, C.J. 63, 73

Reid, W.H. 4–5, 9–10, 13, 72, 77

Risken, H. 35, 73

Robinson, D.W. 18, 73

Rodean, H.C.   .   .   .   .   .   vii, 8, 13, 63, 73–74

Ryzhik, I.M. 57, 72

Sawford, B.L. 4–5, 7–9, 13, 16, 18–19, 24,
       28, 32, 35, 40–41, 45–49, 51–52, 63, 65–66, 71, 74, 78

Schumann, U. 64, 74

Searle, S.R. 36–37, 40, 74, 78

Shao, Y. 68, 74

Shrader-Frechette, K. 69, 73

Skorohod, A.V. 54, 72

Sorbjan, Z. 63, 74

Spanier, J. 57, 74

Sreenivasan, K.R. 66–68, 74

Stankov, B.B. 62–63, 72

Stolovitzky, G. 66–68, 74

Stratonovich, R.L. 26–27, 74

Talkner, P. 18, 72

Taylor, G.I. 4, 9, 74

Tennekes, H. 61, 63, 73

Thomson, D.J. 4–5, 8–10,13, 18–19, 21,
       23–25, 28–29, 31–33, 35–37, 40–41, 43–44, 47, 51–52,
       54, 65, 72, 74–75, 78

Thoroddsen, S.T. 66, 74

Thurtell, G.W. 4, 7, 52, 74–75

Troen, I. 5, 29, 71, 77

Uhlenbeck, G.E.   .   .   .   3–4, 10, 13, 16, 20, 74, 78

Van Atta, C.W. 66, 74

van Dop, H.   5, 9–10, 13, 18–19, 28, 45–47, 51, 71, 74, 78

van Kampen, N.G. 3–4, 7, 10, 13,
       15–16, 18–23, 25, 27, 35, 74, 78

Van Ness, J.W. 67, 73

van Ulden, A.P. 61, 74

Wang, L.-P. 67, 71

Wang, M.C. 3–4, 13, 20, 74, 78

Weil, J.C.   .   .   .   5, 45–49, 52, 54, 61–63, 74–75, 79

Wiener, N. 15, 74

Willis, G.E. 45, 71

Wilson, D.J. 8, 63, 71

Wilson, J.D. 4–5, 7, 9, 18, 25,
       28–29, 33, 44–47, 49, 52–57, 63, 71, 74–75, 79

Wolfram, S. 17, 75

Wyngaard, J.C.   .   .   .   45, 61–63, 71–73, 75, 79

Yaglom, A.M. 8–9, 65, 73, 77

Yee, E. 45–46, 49, 71

Zhu, C.J. 62–63, 72

Subject Index

atmospheric boundary layer . 1, 4, 45, 51–52, 59–64

backward equation 4, 23

boundary conditions 1, 47, 51–57

buoyancy production (or destruction) rate of turbulence kinetic energy 62–63

Borgas solution 40–41

Brownian motion . 1, 3–4, 7, 9–10, 13, 19, 31, 67–69

Chapman–Kolmogorov equation . . 20–21, 52–53

conditional (or transition) probability 4, 15, 19–24, 52–57

convective boundary layer (CBL) 1, 5, 45–49, 59–61, 63

correlation coefficient 9, 67–68

covariance 10, 35–36

cumulants 18

determinant 35–41

deterministic term 7, 25, 32–33, 37–40, 48

diffusion equation (or diffusion equation limit) 5, 28–30, 41–42, 51

diffusion process 13, 15, 17–23

Dirac delta function . . 10, 14–15, 21, 36, 55–56

drift correction (or coefficient) . . . 3–5, 7, 18–19, 24–25, 27–28, 30, 32–33, 37–40, 48

eddy diffusivity (or diffusion coefficient) 9, 28–30, 41–42, 45

Eulerian coordinates 3, 9, 66

Eulerian equation 3, 19, 31

Eulerian turbulence statistics . . 9, 19, 32, 59–64

Eulerian velocity 9, 19, 32, 35, 41

fading memory 7, 25, 32, 37, 48

finite-difference equations 53–57

Fokker–Planck equation 1, 3–4, 15, 18–24, 32, 36, 41, 48

forward equation 4, 23, 32, 36

fractional Brownian motion 1, 67–69

Gaussian distribution . 13–17, 22–24, 33, 35, 45, 55

Gaussian forcing 5, 23, 25, 36, 55

Gaussian process 13–17, 25, 28, 36

Gaussian turbulence . 16, 18, 30, 36, 45, 54, 69

Holtslag–Nieuwstadt atmospheric stability diagram 59–60

Hurst exponent 67–68

incremental Wiener process . . 15–17, 23, 25–30, 31–33, 36, 41–42, 44, 48, 68

inertial subrange 8, 35, 37, 66

intermittency 1, 65–66

inverse of a matrix or tensor . . . 35–36, 43

Ito integral 25–28

jump process 20–22, 24

Kolmogorov's original (1941) hypotheses . 1, 8, 25, 66

Kolmogorov's refined (1962) hypotheses . 1, 66–67

Kolmogorov length and velocity scales . . . 8, 66

Kolmogorov timescale 8, 10, 15, 25, 30, 37

Kramers–Moyal expansion 18, 23–24

Kronecker delta 36–37, 39–40

kurtosis 17, 46–47, 49, 59

Lagrangian coordinates 3, 9

Lagrangian structure function . . 8, 10, 32, 37, 63

Lagrangian turbulence statistics . . . 9, 19, 32

Lagrangian trajectory 1, 4–5, 7, 25, 28, 31–34, 36, 48, 54–57

Lagrangian velocity 1, 4–11, 19, 25, 28–29, 31–33, 35–44, 45–46, 48–49, 52–57

Lagrangian velocity (decorrelation) timescale . . 7, 9–10, 25, 28–30, 32–33, 51–57

Langevin equation (model or LEM) . . 1–11, 19, 25, 28, 31, 33, 36, 44, 48, 54–56

Markov limit 28–30, 51

Markov process . . . 1, 13–17, 20–23, 25, 53, 68

martingale 27–28

master equation 19–22

matrix 35–36

mean (ensemble average) 8–10, 15, 17, 23, 25, 35, 37, 45–46, 48, 55, 59, 65

moments 18–19, 22–24, 46–49

multivariate probability distribution function . 35–36

non-Gaussian forcing 18, 23–24

non-Gaussian turbulence . . . 5, 23–24, 45–49, 69

nonuniqueness 1, 5, 35–41

Ornstein–Uhlenbeck process 4, 16, 18,

parameterization of turbulence statistics 1, 51–52, 59–64

probability distribution (or density) function (PDF) 4, 15, 18–24, 31–33, 35–43, 45–46, 48, 52–57

quadratic form 36

random displacement equation (model or RDM) 1–2, 5, 28–30, 33, 41–42, 54

random (or stochastic) forcing . 3, 5, 18, 23–24, 47–48

random (or stochastic) process . 5, 13–18, 20–22, 25–28

random (or stochastic) term . . . 3, 7, 25, 32, 37

random velocity fluctuation (or increment) . . . 1, 3–11, 19, 25, 28, 31, 35, 44, 48, 54–57

random walk 1, 4, 13–14, 16

rate of dissipation of turbulence kinetic energy . . 8, 10–11, 32, 37, 62–63, 66–67

realizability of a probability distribution function . 1, 18, 23–24, 47

reflection 51–57

Reynolds number 8, 66

Reynolds stress tensor 35–41, 43–44

scaling parameters for the atmospheric boundary layer 59–64

skewness . . . 17, 45–47, 49, 51–52, 54, 59, 63

shear stress 59–61, 64

shear stress production of turbulence kinetic energy 59–61

stochastic differential equation 3, 7, 25–30

Stratonovich integral 26–28

tensor (or tensor notation) 35–44

Thomson's "simplest" solution . . . 40–41, 43–44

transition (or conditional) probability 4, 15, 19–24, 52–57

turbulence statistics 9, 19, 32, 59–64

turbulent diffusion 1, 4–5, 7–11, 25–49

two-particle dispersion 65

unconditional probability distribution (or density) . 22, 32, 35, 48

universal constant of the Lagrangian structure function 8, 10, 32, 37, 63

variance (or velocity variance) 4, 7, 9, 15, 23, 25, 32, 44, 49, 61–64

verification (of models) 69

well-mixed criterion (or condition) . . . 1, 5, 24–25, 29, 31–34, 36, 41, 47–49, 52, 56–57

white noise 14–16

Wiener process 14–17, 67–68